Firm Internal Innovation Contests

W0036973

The Informal Innovation Contests

Björn Höber

Firm Internal Innovation Contests

Work Environment Perceptions and Employees' Participation

With a Foreword by Prof. Dr. Harald von Korflesch

 Springer Gabler

Björn Höber
Koblenz, Germany

Vollständiger Abdruck einer von der Universität Koblenz-Landau genehmigten Dissertation, 2016

ISBN 978-3-658-17491-0   ISBN 978-3-658-17492-7   (eBook)
DOI 10.1007/978-3-658-17492-7

Library of Congress Control Number: 2017933064

Springer Gabler

© Springer Fachmedien Wiesbaden GmbH 2017
This work is subject to copyright. All rights are reserved by the Publisher, whether the whole or part of the material is concerned, specifically the rights of translation, reprinting, reuse of illustrations, recitation, broadcasting, reproduction on microfilms or in any other physical way, and transmission or information storage and retrieval, electronic adaptation, computer software, or by similar or dissimilar methodology now known or hereafter developed.
The use of general descriptive names, registered names, trademarks, service marks, etc. in this publication does not imply, even in the absence of a specific statement, that such names are exempt from the relevant protective laws and regulations and therefore free for general use.
The publisher, the authors and the editors are safe to assume that the advice and information in this book are believed to be true and accurate at the date of publication. Neither the publisher nor the authors or the editors give a warranty, express or implied, with respect to the material contained herein or for any errors or omissions that may have been made. The publisher remains neutral with regard to jurisdictional claims in published maps and institutional affiliations.

Printed on acid-free paper

This Springer Gabler imprint is published by Springer Nature
The registered company is Springer Fachmedien Wiesbaden GmbH
The registered company address is: Abraham-Lincoln-Str. 46, 65189 Wiesbaden, Germany

For my entire family,

their support and encouragement!

Foreword

Innovation Contests as a modern approach for facilitating innovation activities within large corporations are popular and have received increased attention both in theory and in practice. With this dissertation, Bjoern Hoeber enhances the body of knowledge on the integration of firm-internal Innovation Contests into the far-reaching and complex organizational environment of large companies. Here, firms seek to internally conduct such contests to increase its innovation capacity and to exploit its employees' knowledge within the boundaries of the firm.

The overall aims of this dissertation are to investigate the impact of a supportive and facilitative work environment and to identify the necessary areas of action for the design of an engaging work environment for firm internal Innovation Contests. A strong theoretical foundation is provided by both the Componential Theory of Creativity and Innovation in Organizations and the Organizational Support Theory. To investigate this phenomenon in a real-world context, a sequential mixed-methods approach with both quantitative and qualitative parts was chosen.

In the first, quantitative study, the importance of employees' work-environment perceptions (e.g., organizational encouragement and supervisory encouragement of creativity and innovation) as determinants of their individual affective organizational commitment, motivation, and finally their intention to participate in Innovation Contests was proven in cooperation with the product house of a German DAX 30 Company.

In the second, interrelated, qualitative study, the need for diverse areas of action was revealed in light of the experiences and profound expertise of Innovation Contest experts running firm internal Innovation Contests in European companies. As a result, twelve areas of action of an entire organizational support were extracted and characterized.

The findings of both studies offer diverse practical and theoretical implications and help innovation managers to utilize and cope with the challenges of firm internal Innovation Contests. Finally, I would like to strengthen the considerable value of his work and to recommend this book both for practitioners and for researchers.

Prof. Dr. Harald F.O. von Korflesch

Koblenz, Germany

Content

Table of Contents

List of Abbreviations

AC	Affective Commitment
ACS	Affective Commitment Scale
AVE	Average Variance Extracted
CBSEM	Covariance-Based Structural Equation Modeling
CEO	Chief Executive Officer
CFA	Confirmatory Factor Analysis
CFI	Comparative Fit Index
CLF	Common Latent Factor
CMV	Common Method Variance
C.R.	Critical Ratio
CSF	Critical Success Factor
ECS	Enterprise Collaboration Systems
EFA	Exploratory Factor Analysis
EXP	Extra-Role Performance
FFE	Fuzzy front end of Innovation
HR	Human Resources
ICT	Information and Communication technologies
INP	In-Role Performance
KPI	Key Performance Indicator
KMO	Kaiser-Meyer-Olkin-Criteria
LLCI	Lower Level for Confidence Interval
LMX	Leader Member Exchange
ML	Maximum Likelihood
MSA	Measure of Sampling Adequacy
NPD	New Product Development
OE	Organizational Encouragement
OCB	Organizational Citizenship Behavior
OI	Organizational Impediments

OST   Organizational Support Theory

PCA   Principal Components Analysis

PLS   Partial Least Squares

POC   Perceived Organizational Competence

POS   Perceived Organizational Support

PSS   Perceived Supervisor Support

R&D   Research and Development

RBV   Resource Based View

RMSEA   Root Mean Square Error of Approximation

S.D.   Standard Derivation

SEM   Structural Equation Modeling

SE   Supervisory Encouragement

S.E.   Standard Error

SIA   Strategic Innovation Area

SMC   Squared Multiple Correlations

SOE   Supervisor Organizational Embodiment

SRMR   Standardized Root Mean Square Residual

TCE   Transaction Costs Economics

TMT   Top Management Team

ULCI   Upper Level for Confidence Interval

WEI   Work Environment Inventory

WP   Workload Pressure

List of Figures

List of Tables

1 Introduction:

The importance of employees' perceptions for Innovation Contests

This chapter proceeds as follows: First, the motivation for this doctoral thesis (section 1.1) and the derivation of the research questions (section 1.2) are provided. In the next section (section 1.3), the relevance of this work and its intended contributions are presented. Subsequently, the overall research design (methodology, methods, strategy for inquiry, knowledge claims) is introduced (section 1.5). Finally, a summary of the structure of the thesis is provided (section 1.6).

1.1 Motivation

The drivers of this doctoral dissertation are current developments and organizational, instrumental, and socio-emotional changes in the areas of firm internal innovation management, its innovative support using information and communication technologies (ICT), and increased demands for a work environment that supports the innovativeness of large corporations. All three motivational drivers result in various challenges for companies that are briefly presented below.

The first motivation is to address actual developments in the management of innovations within firm boundaries. Over the last ten years, many companies have increased their internal innovation capacity by integrating a wider group of employees into corporate initiatives to develop new products, services or solutions ("company-internal idea generation"). Whereas previously, it was only people affiliated with a firm's research and development (R&D) department who were responsible for all innovation activities (Nobelius, 2004), in the past decade this situation has changed, and now innovations are generated organization-wide (Grote *et al.*, 2012). Therefore, firms are increasingly attempting to activate all its firm internal volunteers to innovate internally (Moeslein, 2013; Haller *et al.*, 2011). A firm's entire workforce can be divided into two large groups that represent sources of innovation: (1) "core inside innovators", who are regularly employed in the R&D department or innovation units; and (2) "peripheral inside innovators". "There is also an interesting [...] and often neglected group of innovators who can be essential for innovation success: peripheral inside innovators. These are employees within the organization who are not directly involved into the innovation process [...] but they innovate mainly due to confidence, curiosity, and pro-active interest in the well-being of the organization" (Moeslein, 2013). More specifically, "employees from all functions, levels, and units of an organization often show extraordinary engagement, motivation, creativity, and talent for innovating" (Moeslein, 2013, p. 70). Additionally, scholars state, "peripheral innovators in terms of employees outside the R&D department have only marginally been taken into account so far. Still, these might provide valuable input, by aggregating all the different views in the organization" (Haller *et al.*, 2011, p. 105). In this setting, Innovation Contests are one efficient method of bringing all interested volunteers together.

Management- and target-oriented innovation development is very important for large corporations because "innovation processes constitute a risk for companies since they require investments with an uncertain result" (Ortt and Smits, 2006, p. 297), particularly because only thirty percent of innovations that are introduced are successful; i.e., there is a "constant failure rate" (Ortt and Smits, 2006, p. 297). Research has highlighted that the use of methods and instruments within a companies' innovation process (for example, Innovation Contests) has a significant and positive influence on large firm's (innovation) success (e.g., Thia *et al.*, 2005; Liedtka, 2011). To bring researchers and practitioners closer to this aim, this thesis mainly focuses on internal assets, namely, the creation of an adjusted work environment and overarching organizational support for Innovation Contests as a firm internal innovation technique.

Second, the use of ICT is widely accepted as a driver for changing intra-organizational processes, as is also the case for R&D, innovation and knowledge-sharing processes (cf. Leimeister *et al.*, 2009; Ebner *et al.*, 2009; McAfee, 2006; Keuper *et al.*, 2013). In particular, the use of social software within companies enables people to collaborate through computer-mediated communication and by forming online communities. Dedicated platforms serve as digital environments for contributions and interactions that are globally visible and persistent over time (McAfee, 2014). Innovation platforms are one prominent example of enterprise social software (McAfee, 2014). With respect to innovation management, McAfee (2009) addresses the potential of enterprise social software tools for transforming companies' innovation processes. Firms "no longer specify who can participate in the innovation process; they welcome all comers" (McAfee, 2009, p. 1). One radical step is that firms not only publicize what they know but also highlight what they need. Traditional and 2.0 approaches for supporting the innovation process reinforce each other (McAfee, 2009). Seeing the innovation process as a firm's most knowledge-intensive organizational process (Adamides and Karacapilidis, 2006), the role of ICT is central in organizing the data and to "support the transformation of information into organizational knowledge" (Adamides and Karacapilidis, 2006, p. 50). The innovation process is strongly collaborative and social and can be supported by information technology for the purposes of collaboration and knowledge management (Adamides and Karacapilidis, 2006).

In recent years, the use of corporate[1] Innovation Contests to support a firm's innovation process has increasingly gained attention both in theory and in practice (Adamczyk *et al.*, 2012). Innovation Contests are especially likely to support the idea generation and selection phases of the innovation process by broadcasting an idea campaign or challenge to a broad group of potential volunteers (cf. Haller *et al.*, 2011). Innovation Contests enrich a firm's knowledge base by integrating contributions and knowledge (Vukovic and Bartolini, 2010). Their strengths are the generation of novel ideas in an efficient and cost-effective manner by using the crowd (cf. Diener and Piller, 2013). Today, Innovation Contests are a frequently realized practical approach (Diener and Piller, 2013; Leimeister *et al.*, 2009) that "has continuously gained in relevance as a corporate innovation practice" (Adamczyk *et al.*, 2010, p. 1). They are assessed as a promising tool with high potential for the firm (Leimeister *et al.*, 2009), an assessment that is also expressed as "effective processes for generating high quality solutions to innovative challenges" (Wooten and Ulrich, 2015, p. 2). To clarify these potential benefits, some examples are provided[2]. Innovation Contests' results include the following: (i) creating 100 ideas in a single day; (ii) creating more than 1,000 ideas in one contest; (iii) reaching 220,000 employees worldwide; and (iv) generating revenues of 1.7 B€ from content created by the firm internal community. Another example involves the positive experiences of AT&T, a global telecommunications provider, underlining the potentials and opportunities of this modern innovation technique. AT&T "has begun to open up its innovation process beyond its labs and to encourage employee participation" (King, 2014). In the last five years, the firm has generated more than 28,000 ideas and activated up to 130,000 members from 54 different countries. AT&T has allocated more than 40 million dollars to find ideas that originated on its online platform, launching more than 75 development projects (cf. King, 2014).

The third motivational driver is that the work environment is important for the innovativeness of corporation: "We know now that the work environment within an organization – which is strongly influenced by management at all levels – can make the difference between the production of new, useful ideas for innovative business growth and the continuance of old, progressively less useful routines" (Amabile, 1997, p. 51). Organizational

---

[1] In this thesis, 'corporate' and 'firm internal' Innovation Contests are used synonymously
[2] Based on the author's personal experiences gained through conducting this research

support and an appropriate work environment have already been the focus of investigations in many contexts, and various positive consequences have been enunciated; *inter alia*, researchers have mentioned positive relationships with employees' affective commitment, job satisfaction and positive mood, interest in the job, and performance in standard and additional job activities (cf. Rhoades and Eisenberger, 2002; Eisenberger *et al.*, 1990; Shore and Wayne, 1993; Shanock and Eisenberger, 2006).

Although both organizational support and its importance for a broad range of intra-organizational activities and processes have already been investigated (for review, see Rhoades and Eisenberger (2002)), a tailored investigation of work environment as a determinant of commitment and innovative behavior in the context of corporate Innovation Contests has been missing. Nevertheless, a few studies focus on various organizational characteristics in a broader sense and their positive influence on Innovation Contest outcomes; e.g., technical and organizational features of the IT platform (Leimeister *et al.*, 2009), autonomy and variety in contest attendance (Zheng *et al.*, 2011), strong and weak ties between participants and contest organizers (Adamczyk *et al.*, 2010) and proactive leadership (Erickson *et al.*, 2012). In summary, because of changes in the highlighted areas, the motivation for this dissertation lies in an in-depth analysis, verification and differentiation of the organizational support phenomenon for influencing employees' desired behavior in an organization-wide innovation development assisted by Innovation Contests as a modern IT platform. A derivation of the research questions is presented in the following section.

1.2 Derivation of the research questions

In the body of scientific literature, an increased number of publications on the emerging topic of Innovation Contests can be recognized, resulting in a research landscape that is manifold and addresses various aspects of the phenomenon (Adamczyk *et al.*, 2012). Several publications have already been published that formulate how Innovation Contests should be designed, managed and supported to archive optimal results. For example, themes such as finding a suitable community (Ebner *et al.*, 2009), the formulation and type of appropriate problems (Terwiesch and Xu, 2008), the evaluation and selection of the best contributions (Yuecesan, 2013) and finding the best technical functionalities for motivating the crowd (Leimeister *et al.*, 2009) are both important and well-understood.

However, scholars such as Argyres and Silverman (2004) have identified an imbalance in the sense that R&D activities are investigated more from an inter-firm perspective (i.e., alliances and networks) than from an intra-firm perspective that considers such activities' influence on the innovation outcome. In addition, Ihl *et al.* (2012 a, p. 1) have noted that "internal management practices, however, that explain why some firms benefit from open innovation more than others are still largely unexplored".

Regarding the state of the art, surprisingly little is known about the influence of organizational encouragement and support on the success of Innovation Contests, although such support is regarded as important: "Innovation Contests have to be managed thoroughly in order to reach the underlying goals. Managing an Innovation Contest incorporates challenging tasks, but essential ones. There are various aspects closely linked to the management if of Innovation Contests such as the motivation of the participants [...] or the support of participants, which organizers of Innovation Contests have to keep in mind" (Adamczyk *et al.*, 2012, p. 344). In this thesis, although the phenomenon of organizational encouragement and support is at the core of all investigations, it is an open field that is not truly understood in the context of corporate Innovation Contests. This dissertation investigates this phenomenon and its influence on employees in real settings in accordance with the following quotation: "The most directly relevant information comes from interviews and survey studies within corporations. It is through these studies that we began to understand the social environment in organizations and how it might impact creativity" (Amabile, 1997, p. 46). Although many research activities could be found in the area of Innovation Contests (for a review, see Adamczyk *et al.* (2012)), this fundamental phenomenon is not completely understood even though their relevance to success might be important: "Although existing research already tackles various issues in regard to the management of Innovation Contests, it is still limited and maintains enough options for further research" (Adamczyk *et al.*, 2012, p. 353).

Thus, this dissertation concentrates on this gap and investigates the phenomenon of the utilization of corporate Innovation Contests throughout a firm. To the best of our knowledge, there has been no previous study that exists to explicitly address this topic. Therefore, the leading research question for this doctoral thesis is as follows:

What is the influence of employees' perceptions on their individual affective organizational commitment, motivation and intention to participate in corporate Innovation Contests and how should the aspects of an engaging work environment be designed?

To answer this question and to close the research gap, this study's aim is to answer the following distinct but related sub-questions:

- **Research question 1: What is the influence of different work environment perceptions on employees' affective organizational commitment and on their motivation and intention to participate in firm internal Innovation Contests?**
- **Research question 2: Which areas of action (critical success factors, constituents and facets) exist that represent aspects of an engaging work environment in firm internal Innovation Contests?**

Given the importance and necessity of organizational support for an optimal utilization of corporate Innovation Contests, the above-described research focus and its relevance could be substantiated from a practical, a theoretical and a scientific-community perspective. Below, the relevance and contributions offered by this doctoral dissertation are described.

1.3   Relevance and contributions

First, this research addresses a practical and highly relevant phenomenon. Both the innovation literature (e.g., Cooper and Kleinschmidt, 1995) and the practical experiences of innovation management software specialists (Al-Ali, 2014) simultaneously hint that organizational encouragement and support is a critical success factor (CSF). It is known that managers and the organization have a "special role to play in helping the Enterprise 2.0 platform take off within their companies" (McAfee, 2006, p. 26) and further, "that use of Enterprise 2.0 technologies is not automatic and depends greatly on decisions made and actions taken by managers" (McAfee, 2006, p. 26). In the context of this thesis, organizational support is defined as all decisions made and actions taken by managers to establish a receptive culture, a common platform, an informal rollout and support by line managers (cf. McAfee, 2006, p. 26).

Nevertheless, firms often struggle with the challenges of the implementation and management of such platforms: "We hear from many innovation professionals that they struggle with their program, often, because the company does not really adopt it. The employees won't share ideas, middle managers are consumed by meeting the quarterly goals, top management sees innovation management as a nice-to-have" (Meisterjahn, 2015). One reason for this problem is that a broad range of aspects must be considered during these complex undertakings. "It is clear that many companies struggle with the cooperative element of innovation. What we have found is that those that have created a culture of cooperation often have done so by developing a cycle of practices and processes that encourage people to believe it is necessary to work together" (Gratton, 2013, p. 107). In addition, challenges often arise in the context of sustainability as highlighted by the following citation: "Research indicates that many visitors to online communities soon disappear" (Bateman *et al.*, 2011, p. 842). On a practical level, the findings help managers, responsible individuals, and organizers of somewhat costly Innovation Contests to increase active participation and to increase both the success and the overall efficiency and effectiveness of such contests.

Theoretically, this dissertation helps management researchers better understand the role and actions of managers supporting the utilization of enterprise social software, especially for innovation management goals. "Since the question of how to manage an Innovation Contest is crucial to its outcome, further research in the management perspective remains pertinent" (Adamczyk *et al.*, 2012, p. 353). This issue is implicated because corporate Innovation Contests are seen as a new type of social information system (Schlagwein *et al.*, 2011) with changed requirements for organizational integration so that new capabilities for supporting corporate Innovation Contests as novel innovation mechanism must be developed. Therefore, excellent companies invest in innovation management capabilities as a form of organizational capability (Lawson and Samson, 2001). These developments are also described in the following quote: "Changing our view of organizations—from focused on command-and-control hierarchies to focused on networks of competency-based virtual communities—promises a radically different set of organizational design options" (Koh *et al.*, 2007, p. 70).

As the groundwork for these large organizational challenges, this dissertation sheds light on work environment perceptions and organizational support as both an important determinant and a critical success factor for Innovation Contests. The executed studies first extend the body of knowledge by conceptually integrating the literature on organizational encouragement and support (the "Componential Theory of Creativity and Innovation in Organizations" and the "Organizational Support Theory" (OST)), motivation and commitment theory, and Innovation Contests. Second, they provide an empirical investigation of the work environment as the determinant of successful Innovation Contests. Third, they provide a qualitative analysis and development of a comprehensive set of success factors and a structured differentiation of work environment aspects tailored to the context of Innovation Contests. Until now, management researchers have paid less attention to the relationship between organizational support by the work environment and employees' motivation and commitment, on the one hand, and intention to participate in such initiatives, on the other hand. Previously, the relevant literature has concentrated on critical success factors for innovation management only on a higher level (cf. Cooper and Kleinschmidt, 2007, 1995).

Finally, this research project responds to several open questions and calls for further research in the innovation management domain. Scholars formulate various unexplored questions that are taken up by this work. Two of them are as follows: "What are adequate organizational structures to support the practice of conducting Innovation Contests?" "What are conditions under which Innovation Contests are most efficient and effective in finding the most innovative ideas?" (Adamczyk *et al.*, 2012, p. 347). As a second call to promote research in this field, a special issue on "organizing crowds and innovation" in strategic organization (Felin *et al.*, 2015) argues that "technological changes have decreased computing and communication costs and transformed the nature of organizational boundaries and the ways firms innovate" (Felin *et al.*, 2015, p. 1). Furthermore, "while there is much work that descriptively captures some of the participatory phenomena and new forms of innovation, there is little work on their theoretical foundations, or their implications for questions such as organizational design, organizational boundaries, leadership and agency, innovation, and the development of organizational capabilities" (Felin *et al.*, 2015, p. 1). Their call addresses several questions, one of which is the following: "How can organizations best engage, utilize, and organize both internal and external participants or crowds when innovating?" (Felin *et al.*, 2015, p. 2). The results of this thesis, especially its demonstration of the broad range of critical success factors for organizational support, clarifies how organizations can facilitate Innovation Contests by paying attention to an ideal work environment as a component of their organizational design.

The domain of politics supports similar themes. For instance, the Federal Ministry of Education and Research in Germany has issued an invitation to tender ideas for models and concepts for the creation of work environments, work organization, governance structures and competences that are inevitable in light of the digitalization of the modern working world (cf. BMBF, 2015). This study serves as a starting point to answer some of the questions opened by these topics.

Next, a content-related positioning of this work is presented.

1.4 A content-related positioning and delimitation

For a clear distinction and to clarify the integration of this work into the research field, a content-related positioning and delimitation of the relevant and widely known topics of "open innovation", "social media", "online communities", "leadership" and "organizational support" and a short assessment of how this thesis is related is set forth below. In addition, an enhancement of the understanding of firm internal Innovation Contests in general might also be accompanied by the information provided in this section.

First, both the innovation literature and the idea of openness or open innovation are jointly considered in many publications (for a review, see Dahlander and Gann (2010)). The journey toward implementing (open) innovation approaches (cf. Chiaroni *et al.*, 2011), which are usually complementary, consists of opening the innovation process both within firm boundaries (e.g., to establish firm-wide exploration networks, integrate firm internal personnel into the innovation process, and leverage a firm's existing knowledge) and beyond firm boundaries (collaborations with either external partners and customers or innovation ecosystems). This thesis investigates "sole" Innovation Contests in the sense of opening up the innovation channel within a firm's boundaries. Accordingly, crowds of innovators outside the firm are outside the scope of this thesis, even if these online communities are also regarded as an important source of innovation for organizations (Franke and Shah, 2003). However, this type of collaboration between volunteers and organizations is fundamentally different than collaboration with employees within firm's boundaries. The differences between external innovation partners and employees are described, e.g., by Sloane (2011).

Second, essential parts of IT-based Innovation Contests are formed by social media and online community features (cf. Leimeister *et al.*, 2009; Bullinger *et al.*, 2010; Hutter *et al.*, 2011; Piller *et al.*, 2012). Today, social media platforms (both firm internal or including members from the external community) are available for almost every group of interests (Kaplan and Haenlein, 2010). Innovation Contests are a special occurrence of a social media platform (McAfee, 2014, 2009). Similar to all other platforms, Innovation Contests support collaboration and the exchange of information between members of the platforms through Web 2.0 functionalities (cf. Leimeister *et al.*, 2009; Bullinger and Moeslein, 2010) "i.e., elements which foster interaction, information exchange, topic related discussion, community building, and—if allowed—collaborative design of products" (Bullinger and Moeslein, 2010, p. 292). Various types of social media target either firm internal or open collaborations beyond firm boundaries. In this work, Innovation Contests as firm internal social media implementation form the basis for all investigations.

In general, online communities "can be seen as a group in which individuals come together around a shared purpose, interest, or goal" and in which "most activity takes the form of posting or viewing options, questions, information, and knowledge" (Koh *et al.*, 2007, p. 70). Additionally, communities are characterized by a "voluntary social context" (Koh *et al.*, 2007, p. 70), and voluntariness is also a primary characteristic of corporate innovation communities (Dumbach, 2014). Within the manner in which Innovation Contests are utilized and implemented as seen in this work, Innovation Contests build and establish a firm internal online community of voluntarily participating employees.

Third, the topic introduced above must be seen in the tension between focusing on leadership (including behaviors by a small group of leaders and direct supervisors in teams, departments, business units, projects) and focusing on organizational support (meaning the body of supporting actions by the entire organization, implying broad features of support from different hierarchical levels and spread over several company members). Both in particular and in general, leadership and organizational support influence individuals' behavior and performance related to creative tasks (Amabile *et al.*, 2004). Nevertheless, the conceptualization and orientation of this thesis classifies it as belonging to the latter end of this spectrum, precisely focusing on an investigation of overarching organizational support and assessing perceptions of the organizational work environment.

Summarizing the content-related delimitation, Innovation Contests are considered as a modern innovation technique that requires a tailored, broad-oriented organizational support for opening the innovation process within firm boundaries, therefore establishing both a firm internal social media platform and virtual community for consolidating innovation efforts.

Below, the research design, and the structure of the thesis are introduced. All these aspects form the methodological foundation for the planned studies and the intended results of the dissertation.

1.5 Research design

In this section, details related to the conceptualization of the investigations and corresponding key elements of the research design, specifically, the "research methodology" (including research methods), the "strategy for inquiry", and the "knowledge claims" (including the paradigms, philosophical assumptions, and epistemology) are presented (cf. Creswell, 2013).

As a research methodology a mixed-methods approach with both qualitative and quantitative methods, following Creswell (2013), is chosen to answer the research questions and to investigate the entire topic. A mixed-methods approach is intentionally chosen because in this case, "the researcher bases the inquiry on the assumption that collecting diverse types of data best provides an understanding of a research problem" (Creswell, 2013, p. 21). The combination of research methods, especially with respect to increasing the robustness of the results, is highlighted in the literature: "In business research, any given objective may require multiple research approaches, often in sequence" (Gable, 1994, p. 3). Gable further states that "through the use of multiple methods the robustness of results can be increased; findings can be strengthened through the cross-validation achieved when different kinds and sources of data converge" (Gable, 1994, p. 4). This dissertation follows a mixed-methods design and in line with the description of Creswell (2013, p. 16), "the study may begin with a quantitative method in which theories or concepts are tested, to be followed by a qualitative method involving detailed exploration with a few cases or individuals". In addition, Kuckartz (2011) highlights the opportunities of a qualitative follow-up phase with the intention of better understanding the results of the quantitative analysis.

Preliminarily, a literature review is provided to increase understanding of both the emerging topic of Innovation Contests and several aspects of their design and management. This research step is based solely on prior knowledge and focuses on analyzing the body of pertinent literature (see section 2.1.3). Next, the following research methods are used within this thesis (see Table 1). This dissertation's research approach is explanatory in the first study (study A) and exploratory in the second study (study B), with the goal of exploring, explaining and predicting a phenomenon in a real-world context, namely, the existence of an engaging work environment that supports the success of Innovation Contests.

Properties	First study (study A)	Second study (study B)
The nature of study	Explanatory, hypothesis testing	Exploratory, inductive reasoning
Study design	Quantitative, causal relationships	Qualitative
Research method(s)	Online survey and statistical ana-	Content analysis

Table 1: Overview of research methodology[3]

First, *quantitative survey research* is chosen both to adapt and verify an existing theory and to test its applicability in the context of corporate Innovation Contests ("theory verification"). Following this aim, study A (with an explanatory study design) empirically investigates different work environment perceptions (both positive and negative, and including organizational support) and those perceptions' influence on employees' *motivation*, on *affective organizational commitment* and on *participation intention*. This study is specially designed to answer research question 1 by using an online questionnaire and the structural equation modeling approach (see chapter 3 for a detailed presentation of study A). To reach this

---

[3] Author's own table.

goal, the "product house" of a large German company in the telecommunications industry that is responsible for group-wide development of the product and service portfolio
offers a rich base for data gathering and statistically testing the relevant relationships.

Second, a *qualitative content analysis* and inductive reasoning are conducted to examine and
understand the phenomenon and subjacent aspects of an engaging work environment
that targets the context of Innovation Contests. This analysis and reasoning is based on
the results of the quantitative investigation. The aim of the qualitative study (study B) is
an identification and in-depth analysis of the areas of action of an engaging work environment in the context of firm internal Innovation Contests (exploratory study design).
Therefore, a cross-case comparison of best practices and a comparison with the relevant
body of knowledge are executed. Case presentations of 18 large, multi-divisional firms
that use corporate Innovation Contests form a rich basis for this analysis. This study is
designed to answer research question 2 (see chapter 5 for detailed presentation of study
B).

Regarding the foci, methods, outcome and contributions of the research steps, different
emphases are given (see Figure 1).

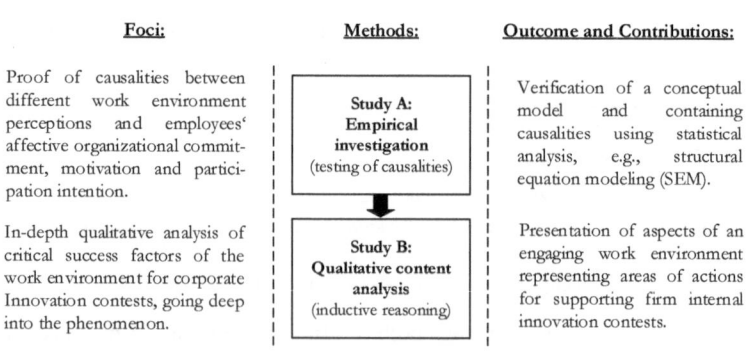

Figure 1: Foci, methods, and outcomes of the research steps in this thesis[4]

More details on the different empirical parts (quantitative study, and qualitative content
analysis) are provided in subsequent chapters.

As a strategy for inquiry, Creswell (2013) differentiates between strategies for purely qualitative, purely quantitative, and mixed-methods research. The strategy for the quantitative
sections of this thesis focuses on survey research as a non-experimental design for data

---

[4] Author's own figure.

collection. After completing the data analysis, a generalization of the findings based on "incorporated causal paths and the identification of the collective strength of multiple variables" (Creswell, 2013, pp. 13–14) is possible because quantitative methods seek "to discover relationships that are common across organizations and hence to provide generalizable statements about the object of study" (Gable, 1994, p. 2). The strategy for the qualitative parts of this thesis could be described according to Creswell (2013, p.15) as "phenomenological research, in which the researcher identifies the 'essence' of human experiences concerning a phenomenon" and "understanding the 'lived experiences' marks phenomenology as a philosophy". Qualitative research seeks "to understand the problem being investigated. It provides the opportunity to ask penetrating questions and to capture the richness of organizational behavior" (Gable, 1994, p. 2). Combining the strategies for the various parts of this research, the strategy for this work is determined as a sequential procedure in which the investigation begins "with a quantitative method in which theories or concepts are tested, to be followed by a qualitative method involving detailed exploration with a few cases or individuals" (Creswell, 2013, p. 16) and in which "a mixed methods design is useful to capture the best of both quantitative and qualitative approaches."

Finally, regarding knowledge claims, purely qualitative approaches often use constructivist or participatory philosophical assumptions, whereas purely quantitative approaches use somewhat postpositivist assumptions (Creswell, 2013). By applying a mixed-methods approach, it is determined that the knowledge claims in this thesis fall within the category of pragmatism, thus, this work is based "on pragmatic grounds" (Creswell, 2013, p. 18) in which "pragmatism opens the door to multiple methods, different worldviews, and different assumptions, as well as to different forms of data collection and analysis." (Creswell, 2013, p. 12). In pragmatic knowledge claims, the problem or phenomenon, its understanding and derivation of knowledge are the most important (Creswell, 2013). Here, and therefore in this thesis, researchers investigate a phenomenon in social contexts, looking at the "how" to reach a specific purpose and focusing on the change to the desired consequences (cf. Creswell, 2013, p. 12) by using both "open-ended observations" and "closed-ended measures". Describing the ontology, which is defined as the assumptions about the nature of reality, scientists (e.g., Klein and Myers, 1999; Bryman, 2012; Creswell, 2013) differentiate between three perspectives, "positivist" (searching for evidence and testing hypothesis, explanatory), "critical" (performing a social critique) and "interpretive" (attempting to understand phenomena through the participants' interpretation). Regarding the studies separately, their ontologies are both positivistic and interpretative in nature. Positivism is present in the quantitative sections in which reality is measured through survey research and a questionnaire (study A). In contrast, the qualitative sections of this research are shaped by interpretivism and realism because there is a focus on the meanings of things that are of importance (study B). Again, combining the ontology of the different parts, the ontology of the mixed-methods approach "is in the middle" (cf. Johnson *et al.*, 2007). Next, and to conclude the introduction, a compact overview of the structure of the thesis and the content of the chapters is presented.

1.6   Structure of the thesis

The structure of the thesis (see Figure 2) is detailed below. This doctoral dissertation starts in the first chapter *("Introduction")* with an overview of its motivation, research questions, relevance and aimed contributions, its content-related positioning, its conceptual framework and research design and a presentation of its structure. The second chapter, *("Theoretical background")*, begins with a review of the state of the art in the body of literature in the areas of intra-organizational innovation and ICT support. Next, a full description and introduction of Innovation Contests, with a focus on intra-organizational implementation, is provided. Chapter 2 closes with the introduction of two relevant organizational theories that form the theoretical foundation for the empirical sections of this work. Next, chapter 3 *("STUDY A: The work environment and participation in Innovation Contests")* depicts the presentation of the first, explanatory study. In the beginning, the study's aim and study and the conceptual model are introduced. More specifically, the selected constructs and their relationships, formulated hypotheses, and the measurement model are depicted. Below, details on data collection, data analysis, a presentation of the empirical findings for study A and the summary are given. In chapter 4 *("STUDY B: Exploring an engaging work environment for Innovation Contests")*, the qualitative content analysis and the results are presented. Therefore, the aim and study design, data collection and analysis procedures, results and a summary are presented in successive sections. Chapter 5 *("Summary, conclusion and outlook")*, contains the summary, conclusion, contributions to research and theory, implications for management and areas for further research.

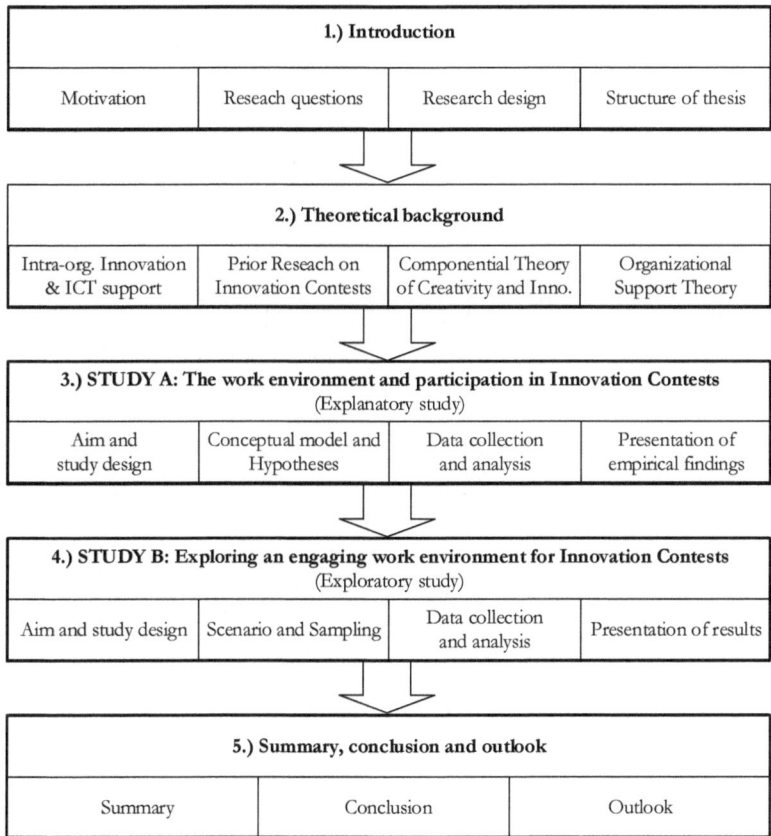

Figure 2: Structure of the thesis

2 Theoretical background: Internal Innovation Contests and the work environment

In the introduction, the positioning of the innovation approach that is investigated in this thesis has been clarified as follows: "Innovation Contests are considered as a modern innovation technique that requires a tailored, broad-oriented organizational support for opening the innovation process within firm boundaries, therefore establishing both a firm internal social media platform and virtual community for consolidating innovation efforts" (Section 1.4, "A content-related positioning and delimitation"). Accordingly, this chapter ("Theoretical Background") first depicts the meaning of intra-organizational innovation (section 2.1.1), the opportunities for support of intra-organizational innovation through the use of information and communication technologies (section 2.1.2) and the state of the art in research on the emerging topic of Innovation Contests (section 2.1.3). Next, two organizational theories are introduced that inform the topic of this thesis and build the theoretical foundation. These theories are the Componential Theory of Creativity and Innovation in Organizations (section 2.2.1) and the Organizational Support Theory (section 2.2.2). Additionally, the "Rubicon Model of Action Phases", which explains individuals' actions and behavior, is presented in section 2.2.3.

2.1 Intra-organizational innovation, ICT support and Innovation Contests

2.1.1 The meanings of intra-organizational innovation

This section introduces important meanings that emphasize intra-organizational innovation efforts within large, often multi-divisional corporations. Furthermore, specialties regarding the fuzzy front end (FFE) of innovation, internal transaction costs, a process for the implementation of novel innovation approaches, and further information on the adoption of (more) open innovation approaches are discussed and presented.

In general, the development of innovation research has an interesting history. The field of (technology and) innovation management (TIM) is already more than 25 years old and "is a multidisciplinary field which continues to grow and evolve over time" (Thongpapanl,

2012, p. 270). The history of various process models for innovations (including technology push, market pull, and the coupling model) and their evolution is illustrated in the generation concept of innovation described by Roy Rothwell (Rothwell, 1994). Baregheh *et al.* (2009) have investigated the concept of innovation in terms of several attributes, including the type of innovation (product, service, process, technical), the nature of innovation (new, change, improve), and the means of innovation (idea, invention, technology, market, creativity).

Addressing the importance and necessity of innovation for a firm's success, "innovation is a basic prerequisite for economic development and the preservation of competitiveness" (Zizlavsky, 2013, p. 1) and "the economics of the 21st century will be characterized by knowledge, information and innovation" (Zizlavsky, 2013, p. 7). Traditionally, innovation activities were executed by a relatively small group of people exclusively or primarily associated with a firm's R&D function (Nobelius, 2004). Collaboration was always face-to-face and focused on products or services. This changed in recent years with a turn toward more modern innovation activities, in which innovations are developed organization wide (Grote *et al.*, 2012). Today, cross-divisional innovation (Grote *et al.*, 2012) or cross-national innovation (Mudambi *et al.*, 2007) require suitable channels for increased knowledge flows with regard to innovative efforts and "the ability of multinational corporations (MNCs) to leverage their innovation competencies across globally dispersed subsidiaries is an increasingly valuable source of competitive advantage" (Mudambi *et al.*, 2007, p. 442). The need for and matter of cross-divisional innovation efforts in large corporations, especially in the early stages of the innovation process, is explained by Grote *et al.* (2012): "In short, the option to recombine highly diversified resources between divisions is very likely to result in considerable value for large corporations. However, [...] that cooperation is difficult to achieve" (Grote *et al.*, 2012, p. 363). The advantages of large corporations over smaller firms are formulated as follows: "The increasing size and diversification of multi-divisional corporations provide these organizations with unique opportunities to discover completely new ways for resource recombination. Compared to arrangements with external organizations, higher secrecy, better access to information and a broader range of available coordination instruments provide additional advantages" (Grote *et al.*, 2012, p. 363).

In the innovation literature, special attention is given to the FFE of innovation (Koen *et al.*, 2004; De Brentani and Reid, 2012). According to Koen *et al.* (2004), the innovation process "may be divided into three areas: The fuzzy front end (FFE), the new product development (NPD) process, and commercialization" (Koen *et al.*, 2004, p. 5). The FFE is characterized by work with an experimental nature, an unpredictable commercialization date, often-uncertain revenue expectations and activities that focus on minimizing risk and optimizing potential (cf. Koen *et al.*, 2004, p. 6). The FFE strongly influences the performance of the entire innovation process because it "is generally regarded as one of the greatest opportunities for improvement of the overall innovation process" (Koen *et al.*, 2004, p. 5). Kim and Wilemon (2002) describe the FFE as "the period between when an opportunity is first considered and when an idea is judged ready for development" that "is one of the most important, difficult challenges facing innovation managers" (Kim and Wilemon, 2002, p. 269). Those authors suggest that subsequent researchers investigate innovative ways to manage the FFE: "What is needed is to develop proper FFE management and practices, which fit situations of the industry" (Kim and Wilemon, 2002, p. 277). It is important to achieve both a balance between creativity and discipline (Khurana and Rosenthal, 1998, p. 59) and a balance between a "pressure of need" and "an environment of open playfulness" (Desouza *et al.*, 2009, p. 11). Khurana and Rosenthal (1998) report that "managing the front end is not easy" (Khurana and Rosenthal, 1998, p. 58) and "successful organizations create a holistic view during the front end, with senior management and core teams adopting a process-oriented style of work, to deliberately link a wide range of technical and organizational considerations concerning business strategy, product decisions, and the subsequent product development project" (Khurana and Rosenthal, 1998, p. 58). Opening up the innovation process within the firm leads to an increase in firm internal networking activities (Kopp, 2011, p. 38). Furthermore, increased networking activities lead to an enhanced exchange of information that is further assessed as source of value: "The patterns of knowledge flows within the firm indicates current sources of value creation and future sources of potential value creation" (Mudambi and Navarra, 2004, p. 388).

The coordination of these activities is one key to success. In this context, internal transaction costs for R&D coordination across units, as investigated by, *inter alia*, Argyres and Silverman (2004), are an important decision factor because these transaction costs strongly influence firms in their decision to choose more centralized or decentralized R&D activities. Internal transaction costs are "costs associated with exchanges or coordination between units or divisions within the firm" (Argyres and Silverman, 2004, p. 933). In the centralized form, a single department is responsible for the firm's research activities. In contrast, the decentralized structure locates research activities exclusively within the divisions. A hybrid form combines elements from both types. Centralization enables a firm to exploit "economics of scale", whereas decentralization is associated with "improved information processing and reduced scope of managerial opportunism" (Argyres and Silverman, 2004, p. 932).

Firms' journey toward the implementation of novel innovation approaches is discussed by Chiaroni *et al.* (2011), who identify a three-phase process that includes the steps of "unfreezing", "moving", and "institutionalizing" (see Figure 3). The unfreezing phase "starts when the firm's top management makes clear its commitment to innovation" (Chiaroni *et al.*, 2011, p. 40) and accordingly, its commitment to the introduction of novel innovation techniques.

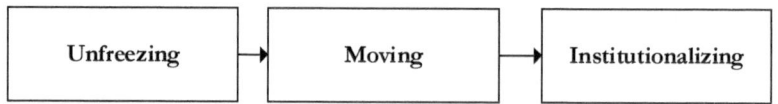

Figure 3: Process steps for implementing a new innovation approach[5]

After the first steps and the development of the first activities, the moving phase is characterized by the "implementation of change [...] that requires an experimental field where the solutions better fitting with the characteristics of the firm are identified" (Chiaroni *et al.*, 2011, p. 41). Finally, through the institutionalizing phase, "the results achieved in the implementation [...] are consolidated and institutionalized" (Chiaroni *et al.*, 2011, p. 41).

Chesbrough and Brunswicker (2014) have asked a large set (n= 125) of firms in Europe and the United States about their adoption of (more) open innovation practices. The results reveal interesting insights. First, 71 percent of the respondents mention increased management support for (more) open innovation approaches compared to 2011 (Chesbrough and Brunswicker, 2014, p. 18). In addition, the intensity, meaning how often and intensively more open innovation practices are being used, "is on the rise" (Chesbrough and Brunswicker, 2014, p. 18). Second, internal employees "were considered the most critical source of innovation ideas" (Chesbrough and Brunswicker, 2014, p. 21). Here, firms continue to regard employees as the key element and "the continued importance of internal employees in the context of an active open innovation program is worth highlighting" (Chesbrough and Brunswicker, 2014, p. 21). Third, managing organizational change and internal cultural issues are mentioned as main challenges and barriers to the introduction of new innovation approaches. The list of benefits enjoyed by firms that use emerging innovation management techniques is relatively long (see Hidalgo and Albors (2008) for a comprehensive overview) and includes "increasing flexibility and effi-

---

[5] Author's own figure.

ciency", "managing knowledge effectively", "increasing productivity and reducing time to market", "facilitating teamwork", "enabling online gathering of marketing information", "improving relationships", "reducing costs by implementing IT-based solutions", and "reducing bureaucratic tasks".

Summarizing the meanings of intra-organizational innovation for large companies, the innovation process and especially the FFE of innovation is very important for companies. Especially larger firms have advantages over smaller firms because of their greater internal workforces available to generate innovations in a firm internal manner. Accordingly, the opening of the innovation process within firm boundaries prompts increased networking and communication activities that might be absorbed by modern innovation approaches. On this occasion, the full introduction of novel innovation approaches and its integration into the corporate environment is characterized by a relatively long journey that involves several phases.

The existential meanings and increased efforts for the development of intra-organizational innovations require support through novel IT-based innovation techniques (Von Kortzfleisch *et al.*, 2015; Moeslein, 2013). Here, ICT and modern social software solutions have a special potential for achieving difficult innovation goals, as presented below.

2.1.2   Intra-organizational innovation and ICT support

This section starts with a presentation of current developments in collaboration systems in general (including the presentation of the concept of organizational embeddedness and the effects on internal transaction costs), followed by the use of modern IT-based tools within a firm, especially for innovation activities. Accordingly, major opportunities for ICT support are described.

In general, solutions for managing and engaging collaboration among employees within large enterprises are widely established. A study on the use of enterprise collaboration systems (ECS) by Williams and Schubert (2015) shows a growing interest in systems that support social interaction. A significant amount of investments in such systems are made by companies. Market growth is estimated from $721 million USD (in 2012) to $6.18 billion USD (in 2018)[6]. The primary benefits, according to the user companies, include the following: Increased knowledge transfer, support for employee collaboration, increased responsiveness, improved expert search, and advancement of innovation and ideas. Con-

---

[6] Williams and Schubert (2015) referencing MarketsandMarkets (2013).

versely, the authors highlight that there remains a high level of uncertainty regarding how ECS should be implemented, used and managed to bring added value to the company.

The effects of computer-assisted communication on organizational design are described in Huber (1990). His conceptual model highlights that the increased availability and use of IT leads to an increase in information accessibility and ultimately to changes in organizational design and improvements in the effectiveness of organizational intelligence and decision making. On the organizational level, centralized organized firms might become more decentralized (and vice versa) in decision-making situations and IT "leads to fewer intermediate human nodes within the organizational information-processing network", meaning more direct communication and a reduction in the number of organizational levels involved. On the subunit level, we see a more intense participation in decision-making processes, a decrease in face-to-face decisions and less time for meetings.

In addition, so-called "social information systems" (Schlagwein *et al.*, 2011) as a novel and emerging group of collaborative information systems are characterized by elements such as web communities, open collaboration, information exchange, and user-generated content. They essentially differ from traditional information systems in dimensions such as sociality, openness, participants and content. Schulze *et al.* (2012a) argue that the use of specialized information systems for idea management leads to higher satisfaction among employees compared to the absence of support through ICT. The implementation of an IT solution can affect both single steps and the entire innovation process.

The use and integration of social software within firm boundaries is illustrated in the literature as the concept of organizational embeddedness (Von Kortzfleisch *et al.*, 2008). In a corporate environment, the use of Web 2.0 "leads to new forms of self-organized, user-centric knowledge and content generation" (Von Kortzfleisch *et al.*, 2008, p. 75). Here, "by using Web 2.0 applications, […] the embeddedness of the actor within the organization is an important aspect" (Von Kortzfleisch *et al.*, 2008, p. 81). The researchers recommend a focus on three core competencies (Von Kortzfleisch *et al.*, 2008) to enhance the organizational embeddedness of Web 2.0 solutions. First, "incentives" must be provided so that they "do not destroy the self-organizing character of Web 2.0 applications" (Von Kortzfleisch *et al.*, 2008, p. 84). Second, regarding the "self-organization" characteristics of such applications and the fact that employees decide autonomously whether they want to participate and how much time they will spend, managers "should recognize that employees will be very sensitive about any attempts to undermine their self-determined behavior" (Von Kortzfleisch *et al.*, 2008, p. 85). Third, with respect to "creativity and innovation", management should enhance "the ability to leave sufficient virtual room for creativity that the self-directed flow of creative ideas remains open, not hindered by management" (Von Kortzfleisch *et al.*, 2008, p. 85).

Technological changes might considerably reduce transaction costs (Kurtmollaiev, 2012) and therefore "conversations and video-conferences are held actually for free. Thus, transaction costs nowadays are much lower than they were in time of shaping TCE[7]" (Kurtmollaiev, 2012, p. 8). Kurtmollaiev explicitly mentions the example of an innovation jam at Volvo Group, where more than one thousand employees have submitted more than 350 ideas and 1,500 comments (Wikhamn and Knights, 2011). He notes that "the sessions, which costs almost nothing for the company now, just twenty years could be unbearably expensive" (Kurtmollaiev, 2012, p. 8). In this context, the "revolution of the time-space intensification" (Wikhamn and Knights, 2011; Odih and Knights, 2002) is described as "increased possibilities to communicate and interact among individuals arguably with less constraints of time, cost and location" (Wikhamn and Knights, 2011, p. 6). Consequently, this approach is "expected to reduce transaction costs at least relating to, [...] searching, evaluating, and monitoring. This in turn would support an intensification of a willingness among organizations to open up their innovation processes" (Wikhamn and Knights, 2011, p. 6).

Focusing on innovation purposes, the value-creation support for innovation through digitization leads to the use of tools and software solutions for the management of firm internal innovation activities (cf. Von Kortzfleisch *et al.*, 2015). Regarding specific ICT support for innovation development, "information technology has, at least to some extent, altered the way that many organisations coordinate and distribute innovation" (Wikhamn and Knights, 2011, p. 5). The use of social software, Web 2.0 technologies and online communities for innovation has become an important strategy for increasing competitive advantage (Di Gangi, P. M. *et al.*, 2010). The characteristics, advantages and disadvantages, obstacles, challenges and success factors of Web 2.0 applications to support innovation activities within the organization are described in several publications (cf. Di Gangi, P. M. *et al.*, 2010; Borowiak and Herrmann, 2011). In general, Web 2.0 applications are assessed to support innovation activities in several ways. For instance, Di Gangi, P. M. *et al.* (2010) mention the elements of knowledge sharing, organizing communication, integrating external partners, and information management as important possibilities. Advantages include a common documentation of knowledge, an asynchronous and multi-

---

[7] TCE= Transaction Costs Economics

local way of collaboration, and the efficiency attributable to rapid knowledge exchange. Challenges might be the creation of freedom for the employees and managing the conflict between leadership and self-organization. In addition, mass acceptance and the recognition of benefits for both individuals and the firm are important, although difficult to achieve.

Moeslein (2013) offers an overview of modern IT-based tools for open innovation approaches (Innovation Contests, Innovation Markets, Innovation Communities, Innovation Toolkits, and Innovation Technologies), whereas "their development, diffusion, and implementation are mainly driven by the attractiveness, usability, and inclusiveness of Web 2.0 and social software" (Moeslein, 2013, p. 72). The author shows that all tools and platform have some interesting common characteristics. The tools commonly (1) allow a large number of innovators, (2) empower collaboration in distributed settings, (3) foster high-speed interactions and (4) build a global memory for innovators.

Below, a brief overview of the tools for open innovation approaches based on the article by Moeslein (2013) is provided[8]. Although Innovation Contests are the selected approach to the investigations in this thesis, the presentation of other, similar approaches helps to show the benefits and bandwidth of ICT support for Innovation activities. First, "Innovation Contests" often are defined as web-based competition among innovators to solve a particular challenge (Moeslein, 2013; Adamczyk et al., 2010). Web 2.0 and social software components enable a global reach with low costs. Innovation Contests are implemented in various ways, are often time limited, and are characterized by a high level of interaction to foster both creativity and idea quality. Next, "Innovation Markets" "act as intermediaries, connecting innovation seekers and innovation providers (often called "solvers")" (Moeslein, 2013, p. 74). Prominent examples of professional Innovation Markets platforms include "Atizo", "InnoCentive", and "NineSigma". Additionally, "Innovation Communities" "enable innovators to collectively share and develop ideas, discuss concepts, and promote innovations. Web 2.0 and social software based innovation communities normally bundle interested and specialized innovators for particular issues and thus support collective development and enhancement of innovation concepts" (Moeslein, 2013, p. 76). Prominent examples include open source development communities, e.g., the "Apple Developer Connection" community. In innovation communities, volunteers do not compete directly with each other. "Innovation Toolkits" "provide an environment

---

[8] cf. Moeslein (2013, pp. 72ff.) for detailed information.

in which users develop solutions in prescribed steps. [...] The application of toolkits is quite widespread for the configuration of predefined solutions, the mass-customization of predesigned products, and the selection of variants from a broad range of offerings" (Moeslein, 2013, p. 77). Prominent examples include toolkits for customizing cars (often called "configurators"). Finally, various "Innovation Technologies" "like 3D-scanners, laser cutters, or 3D-printers allow even individual users to fully develop new products and services" (Moeslein, 2013, p. 79). Furthermore, they "enable progress from the conceptualization of an innovation to prototyping or even producing a product or service" through an integration with social media platforms.

Summarizing the given information in this section, pertinent literature highlights the recently increased use of collaboration systems in general and the use of such systems for innovation purposes in particular. Here, social software systems and diverse modern IT tools for innovation offer different opportunities for support. However, and to ensure the success of the implementation, these modern approaches must be integrated and embedded into the specific organizational context.

In the following section, the state of the art in prior research on the topic of Innovation Contests as the selected approach for this thesis is presented.

2.1.3 Prior research on Innovation Contests
In the previous sections, the meaning of intra-organizational innovation and the opportunities for support through ICT and social media are presented; moreover, diverse tools for (open) innovation purposes are discussed. I focus exclusively on Innovation Contests as the selected approach for this thesis's investigation because of its wide dissemination, as detailed below. Therefore, an overview of Innovation Contests (section 2.1.3.1), central aspects of the design of Innovation Contests (section 2.1.3.2) and selected further research (section 2.1.3.3) are presented.

2.1.3.1 An overview of Innovation Contests
In recent years, intensified attention has been paid to Innovation Contests both in practice and in research. The wide dissemination of Innovation Contests in practical use is highlighted in the scientific literature because "an increasing number of organizations worldwide have adopted Innovation Contests" (Adamczyk et al., 2012, p. 335). Simultaneously, "Innovation Contests represent a growing research field to scholars from different backgrounds" (Adamczyk et al., 2012, p. 335).

The pertinent literature predominantly relies on the term "Innovation Contest" instead of "idea contest" "to illustrate that a contest is able and suited to cover the entire innovation process from idea creation and concept generation to selection and implementation" (Hallerstede and Bullinger, 2010, p. 2). Nevertheless, several terms are in use, including "innovation competition", "ideas competition", "design contest" and "design competition" (cf. Adamczyk et al., 2012, p. 339). A comprehensive literature review[9] on the topic of Innovation Contests is provided in Adamczyk et al. (2012). Those authors define Innovation Contests "*as IT-based and time-limited competitions arranged by an organization or individual calling on the general public or a specific target group to make use of their expertise, skills or creativity in order to submit a solution for a particular task previously defined by the organizer who strives for an innovative solution*" (Adamczyk et al., 2012, p. 335). Additionally, there is a previous definition of Innovation Contests as a "*web-based competition of innovators who use their skills, experiences and creativity to provide a solution for a particular contest challenge defined by an organizer*" (Adamczyk et al., 2010, p. 3). Alternatively, Blohm et al. (2011a, p. 2) highlight that "an idea competition can be defined as an invitation of an organizer, namely a firm, to the general public or a targeted group to submit contributions to a certain topic within a timeline. An idea reviewers committee evaluates these contributions and selects the winner".

Regarding the state of the art, a broad distribution of publications can be seen: "Innovation Contests are a topic for researchers of various subject areas. Publications on Innovation Contests are scattered across diverse journals and conferences in the different subject areas" (Adamczyk et al., 2012, p. 339). In their literature review, Adamczyk et al. (2012) identify publications from various research areas, namely, Management Science, Economics, Computer Science and Information Systems, Education Science, and Sustainability (Adamczyk et al., 2012, pp. 339 f.).

The basic functionality of the Innovation Contest approach is described as follows: "By organizing an Innovation Contest, the pool for idea generation […] is broadened as the organizer taps into the wisdom of the crowd […]. Submissions, i.e., ideas/designs […] or even concepts and solutions […], are used as creative input for the development of new products and services" (Haller et al., 2011, p. 104). Thus, the successful design and management of Innovation Contests might be a complex issue because several determinants influence how Innovation Contests could be handled and conducted. "A lot of different

---

[9] An earlier version of this literature review is provided by Bullinger and Moeslein (2010).

tasks linked to the management of Innovation Contests are included" (Adamczyk *et al.*, 2012, p. 343).

Subsequently, selected central aspects of the design of Innovation Contests are provided to further increase the understanding of the Innovation Contest approach.

2.1.3.2 Central aspects of the design of Innovation Contests

Because of its wide field of application, diverse aspects of the design of Innovation Contests must be considered; these aspects can vary widely over the existing implementations. The pertinent literature addresses the following issues in particular: (1) various **perspectives** on Innovation Contests (e.g., Adamczyk *et al.*, 2012); (2) the **contestants** and their behavior (e.g., Neyer *et al.*, 2009; Hallerstede and Bullinger, 2010; Boudreau *et al.*, 2011); (3) **examples** from industries (e.g., Hallerstede and Bullinger, 2010; Bayus, 2013; Bjelland and Wood, 2008); (4) recurring **design elements** (e.g., Bullinger and Moeslein, 2010; Adamczyk *et al.*, 2012); and (5) the **process** and responsibilities involved in an Innovation Contest (e.g., Haller *et al.*, 2011; Bullinger *et al.*, 2010). These issues are detailed below.

First, by illustrating different **perspectives** on Innovation Contests, Adamczyk *et al.* (2012) review and summarize scientific publications on Innovation Contests from an "economic perspective" and a "management perspective" and by using different foci (see Figure 4). The economic perspective provides "a deep understanding of the economic model behind Innovation Contests" (Adamczyk *et al.*, 2012, p. 342). *Inter alia*, topics such as the effort and costs incurred by the organizer and the contestants or questions about the prizes are discussed within this perspective. Conversely, the management perspective addresses "certain aspects in regard to the management of Innovation Contests" and "strives to understand how Innovation Contests could be handled and conducted" (Adamczyk *et al.*, 2012, p. 342). In addition to these two general perspectives, the focus of Innovation Contests can relate to education, innovation or sustainability. First, the "education focus" has "the primary purpose of encouraging and motivating students to develop technical, design, teamwork, and communication skills" (Adamczyk *et al.*, 2012, p. 343). Second, the "innovation focus" of Innovation Contests is described as "the usage of Innovation Contests to stimulate and foster the development of new products or services in order to achieve innovation objectives" (Adamczyk *et al.*, 2012, p. 345). Third, the "sustainability focus" considers "Innovation Contests that are conducted with the primary aim of mastering sustainability issues" (Adamczyk *et al.*, 2012, p. 346). Applying the perspectives to the topic at hand, this thesis proceeds according to the management perspective of Innovation Contests (in addressing how to handle and to conduct firm internal Innovation Contests) and takes an innovation focus (to foster the development and improvement of products and/or services).

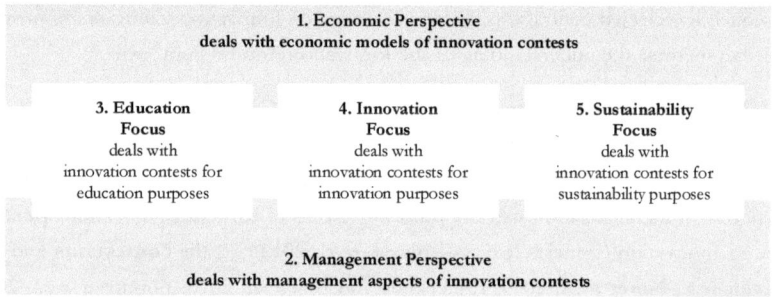

Figure 4: Various perspectives and foci of Innovation Contests[10]

Second, regarding **contestants** in general, "Innovation Contests are suitable tools for the integration of various stakeholders into the innovation process" (Adamczyk *et al.*, 2012, p. 346). More specific, Neyer *et al.* (2009) differentiate three groups of innovators from inside and outside the organization, namely, the "core inside innovators", the "peripheral inside innovators", and the "outside innovators". A description of these groups of innovators is presented in Table 2.

---

[10] Author's own figure, referring to Adamczyk *et al.* (2012).

Group of Innovators	Description
Core inside innovators	*"The core inside innovators, i.e., the R&D departments are traditionally held responsible for the generation and collection of innovative ideas within or beyond the boundaries of the organization."*
Peripheral inside innovators	*"The peripheral inside innovators are employees across all business units within the organization. From the perspective of the R&D department, they sit at the periphery. By their daily work, however, the peripheral inside innovators become knowledgeable and involved experts. This group is not responsible for innovative activity by their job description, but nonetheless is interested in and has the potential to produce innovative ideas and contribute to the innovation process by suggesting, supporting, or refining innovative concepts."*
Outside innovators	*"The outside innovators, i.e., the entirety of external partners—(end) customers, users, retailers, suppliers, and competitors—can act as initiators and/or participants of the innovation process as well. Multidisciplinary participants ensure that the different needs and capabilities as well as the integrated perspectives of market and technology can positively influence the innovation process from the beginning."*

Table 2: Groups of innovators[11]

In the case of focusing on opening up the innovation process at an intra-organizational level, Vujovic and Ulhoi (2008, p. 153) call this strategy as "the boundedly open innovation strategy". The approach of using the labor of the entire internal staff could be compared to the "inner sourcing" practice of the field of open source development. Inner sourcing is referred to as such because "the software is sourced internally" and furthermore, "an inner-source project does not belong to a single team or department. Anybody in the organization can be a contributing member of the community" (Stol and Fitzgerald, 2015, p. 60).

In contrast, focusing on opening up the innovation process at an inter-organizational level, Hutter *et al.* (2011, p. 3) highlight this strategy as "following the concepts of crowdsourcing, co-creation or open innovation, companies are increasingly using contests to foster the generation of creative solutions." As described in the motivation and intro-

---

[11] Author's own table, direct quotations from Neyer *et al.* (2009, pp. 410 f.).

duction sections of this thesis, a specialization in firm internal Innovation Contests focus-es on peripheral inside innovators and consonant with a boundedly open innovation strategy.

For contestants' behavior in terms of collaboration, cooperation, and competition, the principle of cooperative orientation in Innovation Contests (Bullinger *et al.*, 2010; Obstfeld, 2005) describes a continuum (see Figure 5) anchored at a low cooperative orientation (corresponding to competition behavior between submissions) and a high level of cooperative orientation (corresponding to cooperation behavior between submissions).

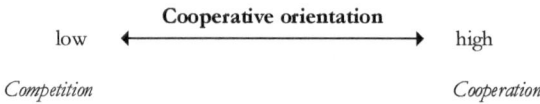

Figure 5: The continuum of cooperative orientation[12]

Following Bullinger *et al.* (2010, pp. 293 f.), individuals with high cooperative orientation "have contacts outside their social group", "introduce disconnected individuals" and "facilitate cooperation". Conversely, people with a low cooperative orientation "act as lone fighters" and "perceive the environment as competitive". Boudreau *et al.* (2011, p. 1) highlight that "greater rivalry reduces the incentives of all competitors in a contest to exert effort and make investments. At the same time, adding competitors increases the likelihood that at least one competitor will find an extreme-value solution". The importance of providing new ideas and giving feedback is addressed in the literature (e.g., Adamczyk *et al.*, 2011a; Adamczyk *et al.*, 2011b). "Listening to comments by other users can even overcome a lack of individual knowledge" (Adamczyk *et al.*, 2011b, p. 221). Therefore, "these platforms allow users to submit ideas and use community functionalities to mutually comment and evaluate their activities. Both submissions and comments by idea owners, community managers, and peers hold considerable knowledge" (Adamczyk *et al.*, 2011a, p. 232).

To take advantages created by this opportunity, "community functionality" is "essential to keep members on the platform and to establish an active community" (Adamczyk *et al.*, 2011a, p. 234) and "needs to be carefully designed and implemented" (Adamczyk *et al.*, 2011b, p. 221). In a qualitative analysis, Adamczyk *et al.* (2011a, pp. 238 ff.) have discovered 22 different types of comments, including "sharing experience/information", "con-

---

[12] Author's own figure, referencing Bullinger *et al.* (2010, p. 296).

fessing problems", "positively evaluating submissions", "asking questions", and "sharing opinions".

In terms of the various levels of cooperation, the literature distinguishes three different forms of cooperation in Innovation Contests (cf. Digmayer and Jakobs, 2013; Hallerstede and Bullinger, 2010). First, "community-based contests" are characterized by a hedonic community, the use of community functions, non-monetary rewards, a short running time, and a medium level of elaboration of contributions. Second, "expert-based contests" invite qualified professionals in the relevant subject area and are characterized by the use of community functions, monetary rewards, long running time, and a high level of elaboration of contributions. Third, "mob-based contests" usually do not enable collaboration and do not use community functions. Instead, participants only submit their contributions to the platform. A separate categorization is provided by Terwiesch and Xu (2008), who differentiate between expertise-based projects, ideation projects, and trial-and-error projects. According to Terwiesch and Xu (2008, p. 1533), "expertise-based projects" are characterized by "engineering tasks with no uncertainty in performance function (well behaved solution landscape)". "Ideation projects" focus on "innovative problems with no clear specifications, leading to uncertainty in the performance function". Finally, "trial-and-error projects" attempt to find "solutions to research problems with well-defined goals, yet highly rugged solution landscapes, creating uncertainty in how to improve a solution".

Third, **examples** of Innovation Contests organized by firms in different industries are available in different publications. With respect to practical utilization, the broad dissemination of Innovation Contests is highlighted, for example, as follows: "Nowadays, with the global availability of broadband access to the World Wide Web, IT-based Innovation Contests are used for a broad range of tasks—from designing wristbands for watches (e.g., Swarovski) to solving complex scientific problems (e.g., XPrize foundation)" (Haller *et al.*, 2011, p. 103) Bullinger *et al.* (2010, p. 292) illustrate "the rich diversity of Innovation Contests in practice". For instance, Adamczyk *et al.* (2012) (referencing BMW, IBM, and Siemens) or Hallerstede and Bullinger (2010) (referencing 53 Innovation Contests, including those of Samsung, Lego, Google, and Vodafone) offer examples of online Innovation Contests. Firms as organizers of online platforms for customer integration in the innovation process (not explicitly using contests) are described by Rohrbeck *et al.* (2008) (referencing, *inter alia*, Allianz, Deutsche Telekom, Microsoft, and Procter & Gamble). Some exemplary initiatives are depicted in Table 3.

Firms	Initiatives	References	Aim of Initiative
ADIDAS	*Adidas idea competition*	*Piller and Walcher (2006)*	*In a research project, the members developed an idea competition together with the firm and launched it in a selected market.*
CISCO	*Cisco I-Prize*	*Jouret (2009)*	*With 2,500 innovators, Cisco aims to find a new idea for a billion-dollar business.*
DELL	*Dell's idea storm*	*Bayus (2013)*	*IdeaStorm's goal is to give customers the opportunity to suggest new products or services.*
IBM	*IBM's innovation jam*	*Bjelland and Wood (2008)*	*IBM's innovation jam was used to bring employees and stakeholders together to move the company's technologies to market.*
OSRAM	*OSRAM LED design contest*	*Hutter et al. (2011)*	*OSRAM designed new LED Light Solutions with its customers in a community-oriented idea contest.*

Table 3: Examples of firm-organized Innovation Contests[13]

Forth, Adamczyk *et al.* (2012) and Bullinger and Moeslein (2010) describe a set of ten different recurring **design elements** for Innovation Contests (see Figure 6). They provide a holistic systematization for the design and configuration of Innovation Contests and how they could be organized; "design elements and their attitudes can influence the appearance of innovation contests, which might be very different. […] In order to align innovation contests with their purposes, […] design elements should be chosen correspondently" (Adamczyk *et al.*, 2012, p. 349). A detailed introduction of these design elements is not given here; instead, I refer to Adamczyk *et al.* (2012, pp. 348 ff.).

---

[13] Author's own table.

Design elements	Attributes				
1 Media	Online		Mixed		Offline
2 Organizer	Company	Public organization	Non-profit		Individual
3 Task/Topic specifity	Low (Open task)		Defined		High (Specific task)
4 Degree of elaboration	Idea	Sketch	Concept	Prototype	Solution	Evolving
5 Target group	Specified			Unspecified	
6 Participation as	Individual		Team		Both
7 Contest period	Very short term	Short term	Long term		Very long term
8 Reward/Motivation	Monetary		Non-monetary		Mixed
9 Community functionality	Given			Not given	
10 Evaluation	Jury evaluation	Peer review	Self assessment		Mixed

Figure 6: Design elements of Innovation Contests[14]

Fifth, with respect to the **process** and responsibilities of Innovation Contests, several publications offer interesting insights (cf. Haller *et al.*, 2011; Bullinger *et al.*, 2010). A rather simple differentiation of responsibilities is available; Innovation Contests "basically consist of two distinct activities for participants: (1) submitting ideas related to a specific topic and (2) evaluating, commenting and improving other participants' ideas. The former activity is typically carried out by individual users, whereas the latter is a more collectively oriented task where users interact and assist each other" (Ihl *et al.*, 2012b, p. 1). A more nuanced representation (see Figure 7) is given by Haller *et al.* (2011, p. 103). Following their description, the Innovation Contest process starts with the step of "publishing the challenge" by the organizer. Afterwards, the participants are responsible for the "submission of contributions", for the "improvement of contributions" and for the "community evaluation". Finally, the organizer coordinates the "jury evaluation" and the "rewarding" of the winning submissions.

---

[14] Author's own Figure, referencing Adamczyk *et al.* (2012).

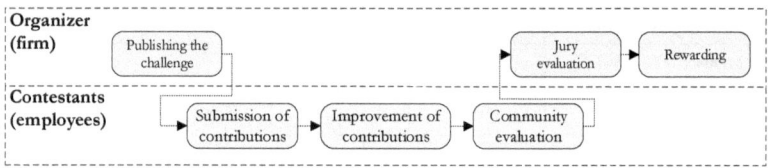

Figure 7: The Innovation Contest process[15]

Table 4 offers further details on the process of Innovation Contests by mentioning the main information objects and a brief description for every process step.

Process steps	Information objects	Description
Publishing the challenge	*Problem Statement*	*The organizer publishes the open challenge online on the platform.*
Submission of contributions	*Ideas*	*Participants submit their contributions during a defined time period.*
Improvement of contributions	*Comments*	*Participants elaborate on contributions during a defined time period.*
Community evaluation	*Votings*	*Contributions are evaluated and pre-selected by the community.*
Jury evaluation	*Evaluations*	*Contributions are assessed and ranked by a jury of experts.*
Rewarding	*Incentives*	*Winning submissions are rewarded (monetary/non-monetary prizes).*

Table 4: Steps of the Innovation Contest process[16]

Until now, the central aspects (perspectives, contestants, examples, design elements, and the process) of the design of Innovation Contests have been highlighted. Next, selected further research on Innovation Contests is briefly introduced.

2.1.3.3 Selected further research on Innovation Contests

In addition to the presented overview and the central aspects of the design of Innovation Contests as presented on the preceding pages, there is a body of literature that further addresses aspects of Innovation Contests. The majority of the available and reviewed publications focus on Innovation Contests as an approach to inter-organizational collabo-

---

[15] Author's own picture, adapted from Haller *et al.* (2011).
[16] Author's own table, referencing Haller *et al.* (2011).

ration that integrates diverse external partners into the firm's innovation process. Nevertheless, this literature can offer interesting insights, as presented below along the process steps of Innovation Contests set forth above.

In step A, "publishing the challenge", the formulation of a precise depicted task is an especially important determinant of the success of Innovation Contests: "Formulating the innovation challenge for the contest is crucial" (Moeslein, 2013, p. 74). This step primarily influences the number of participants (Yang *et al.*, 2009). In addition, the challenge must be appropriate to the addressed community. Terwiesch and Xu (2008) have analyzed which types of problems can be solved using Innovation Contests. In general, Innovation Contests might not be an appropriate technique if market and technological uncertainty is simultaneously high. Afuah and Tucci (2012) investigate the use of online contests from a problem-solving perspective. Tasks should be clearly delimitable, simply transferable and characterized by a small amount of implicit knowledge and complexity. They introduced so-called solution landscapes and differentiated local from more distant solutions. The knowledge required to solve the problem should have a large effective distance from the knowledge base of the solution seeker. Otherwise, Innovation Contests are not suitable. Contests are the preferred mechanisms if the solution (not the task) is characterized by a high level of implicit knowledge and high complexity. Katila and Ahuja (2002) investigate similar approaches to the problem of local search ("local search bias"). Schulze *et al.* (2012b) postulate that the demands-abilities fit, meaning the conformity between the "requirements of the task" and "individual competencies", must be high. Furthermore, the needs-supplies-fit, meaning the conformity between the "properties of the task" and "individual needs", must be equally high. The levels of these fits are important determinants for the intention of individuals to participate.

In step B, the "submission of contributions", the pertinent publications address, *inter alia*, the size of the crowd, individuals' motives for participation, and fairness expectations are important determinants. Regarding the size of the crowd, one strategic decision is how many individuals should participate in Innovation Contests, as highlighted below. It is important for organizers to increase the amount of active participants because employees' active participation of employees is critical to success (Leimeister *et al.*, 2009). Research has already investigated several positive consequences of active participation. Terwiesch and Xu (2008) mention contests in the interplay between a higher or lower number of participants. Whereas a larger group of people increases interdisciplinarity, a small group rather ensures individual effort. Boudreau *et al.* (2011) analyze the impact of the total number of participants on performance. Different effects could be observed in the case of an increase of participants. First, the incentive effect describes a decrease in individuals' personal effort, leading to a lower overall effort. The "parallel path" effect describes a broadening of published ideas and a broader search for suitable solutions.

Participation is primarily influenced by the activation and motivation of the participants (Leimeister et al., 2009; Frey et al., 2011). However, "being motivated implies a number of different phenomena" (Gollwitzer, 1990, p. 53). Various motives that influence individuals' propensity to participate in crowdsourcing contests are known. For example, incentives and rewards as antecedents for motivation have been investigated (Cahalane et al., 2013). Four motives that influence individuals to participate in online contests can be distinguished (Leimeister et al., 2009). These motives include direct compensation (e.g., financial rewards), social motives (e.g., appreciation by the community), self-marketing (e.g., profiling) and learning (e.g., access to new knowledge). Other research relies on the distinction between extrinsic and intrinsic motivations (Frey et al., 2011). In a firm-hosted community, users freely reveal ideas because of their individual attributes (Jeppesen and Frederiksen, 2006). These attributes positively affect their willingness to share innovative ideas. One motivating factor is the possibility of recognition by the firm. One study reveals that user-generated ideas could be significantly higher in terms of novelty and customer benefit in compared to ideas generated by a firm's professionals (Poetz and Schreier, 2012).

Participants' fairness expectations are primarily generated based on the "terms and conditions" and ex-ante identification with the organizing firm (Franke et al., 2013). These expectations influence the individuals' decision about whether to participate in such contests. In line with Franke and colleagues, fairness could be divided into distributive and procedural fairness. First, the more clearly regulations regarding monetary benefits, reputation and intellectual property favor the organization, the more negative are individuals' expectations of distributive fairness. Second, the less transparent the regulations regarding the process and selection criteria are, the lower are the individuals' expectations of procedural fairness. A low personal identification with the firm strengthens negative expectations and leads to a lower degree of willingness to participate. In general, a positive knowledge-sharing behavior is influenced by the perception of fairness (Roetzel and Lohmann, 2014).

In the next step (step C) of the Innovation Contest process, "improvement of contributions" is often highlighted as important (cf. Adamczyk et al., 2011a; Adamczyk et al., 2011b); "Feedback from other users helped the ideas presenters to refine their ideas during the runtime" (Leimeister *et al.*, 2009, p. 208). Hutter *et al.* (2011) postulate a simultaneous encouragement of competitive and cooperative behavior. A competitive orientation within the crowd supports the creation and formulation of high-level contributions. Competition could be encouraged by individually allocated incentives. Cooperative behavior supports the exchange of knowledge between individuals or teams. Cooperation could be encouraged by incentives for group contributions or publishing feedback or comments. The duality of competition and cooperation in Innovation Contests has also been investigated by Bullinger *et al.* (2010). Findings suggest that either a high or a low level (but not a medium level) of cooperative orientation can lead to a high level of innovativeness.

On a related note, authors have addressed the question of how many rounds should be included in contests. Yu and Nickerson (2011) suggest a multi-phase process for Innovation Contests. Here, ideas developed by one group of participants are the starting point for further processing by separate communities. A multi-phase process increases the creativity, originality and feasibility of submitted ideas. Terwiesch and Xu (2008) also suggest studying whether contests should be executed in several rounds.

Additionally, a combination of online and offline activities helps to foster interaction and collaboration. Jung *et al.* (2012) have developed an integrated approach that includes different activities. These activities foster intensive discussions among participants. The authors suggest discussion rounds to understand and discuss the problem, assumptions and a combination of ideas. Positive effects include the maximization of benefits and solutions that have a higher degree of innovation.

With respect to step D, the "community evaluation", the selection of the best ideas is a very important, but difficult, step. Yuecesan (2013) argues that the strength of Innovation Contests is more than the generation of novel ideas. However, although the screening- and evaluation process is often costly and error-prone, "the integration of larger (external) groups in the evaluation of innovations seems to generate better results" (Bullinger and Moeslein, 2010, p. 5). Evaluation decisions often must be made quickly and on the basis of incomplete information (Yuecesan, 2013). Because a firm's resources to evaluate ideas are limited, "there is a strong need for mechanisms supporting the evaluation of these ideas" (Blohm *et al.*, 2011c, p. 1). Here, using the community evaluation as "mechanisms for community-based idea evaluation may facilitate the process of identifying the 'best' ideas" (Blohm *et al.*, 2011c, p. 2). Community evaluation should help in meeting the challenge of evaluating and selecting the best contributions, a process that has been assessed as a "resource intensive" activity (Fueller *et al.*, 2010).

In this context, the concept of open evaluation is described as scenario that "represents and bundle the judgment of people who are not part of the general group of decision makers" (Fueller *et al.*, 2010, p. 956). "Recent research indicates that open evaluation bears plenty of potential to support the selection of relevant submissions" (Fueller *et al.*, 2010, p. 955). Here, the organizers "rely on the opinion of the participants to pre-filter relevant submissions. They provide active feedback [...] and their preferences by voting for or against certain submissions. Even detailed evaluations can take place" (Fueller *et al.*, 2010, p. 956). Riedl *et al.* (2010, p. 2) describe this approach as the "collective decision making of many individual evaluations". Those researchers show that more complex, multi-attribute scales outperform single rating mechanisms such as thumbs up/down and 5-star rankings. Blohm and his colleagues compare the use of rating scales and prediction markets (so-called idea or innovation stocks) in innovation communities and prove that a "multi-criteria rating scale outperforms prediction markets in terms of evaluation accuracy and evaluation satisfaction" (Blohm *et al.*, 2011c, p. 1). Finally, the community-evaluation approach can lead to so-called evaluation games for finding promising contributions in Innovation Contests (cf. Fueller *et al.*, 2010). Here, the "players" are randomly matched and asked to compare two submissions from a large set to evaluate which one they prefer ("pair-wise comparison"). If their evaluation matches, they earn points. Based upon this approach, a general ranking of all submissions can be built.

Step E, the "jury evaluation", is a significant step: "The use of a jury is probably the most longstanding and most widespread evaluation approach in Innovation Contests", and "jury members usually show a certain expertise, experience, or position so that they can act as process or power promotors for the implementation and/or commercialization of the winning solution" (Fueller *et al.*, 2010, p. 956). Bullinger *et al.* (2010, p. 292) state that "analysis of practice shows that the winning submission is predominantly selected by a jury of experts". For the jury evaluation, the "user evaluations were a helpful measure in the later evaluation phase when the submissions were evaluated and discussed" (Leimeister *et al.*, 2009, p. 208), thus "the evaluations conducted by the community served as a filter mechanism for a jury of experts to select the most interesting ideas" (Hutter *et al.*, 2011, p. 6). The consensual assessment technique (Amabile, 1996) is a suitable mechanism (cf. Piller and Walcher, 2006; Blohm *et al.*, 2011c) for evaluating innovative ideas. According to this method, "products or responses are creative to the extent that appropriate observers agree that they are creative. In this context, appropriate observers are people who are familiar with a domain" (cf. Amabile, 1996, p. 4).

Generally, the evaluation of contributions in Innovation Contests is performed based on predefined evaluation criteria, whereas "evaluation criteria in Innovation Contests are manifold, including frequent assessment of the creativity and workability of ideas and designs but also their aesthetics" (Fueller *et al.*, 2010, p. 956). One set of possible evaluation criteria for idea competitions is highlighted by Ebner *et al.* (2009, p. 350). They define a set of evaluation criteria that considers various dimensions of submitted ideas (see Table 5). To identify the best ideas, the overall score "is calculated by the average of weighted grades (from '1= strong agreement' to '5= strong disagreement') assigned to each dimension of the evaluation criteria" (Ebner *et al.*, 2009, p. 349).

Dimensions	Criteria	Description
Creativity	*Originality*	*The degree in which the idea is novel and unique*
	Degree of Innovation	*The idea is a new combination of factors, which can be utilized for economic benefit*
Market potential	*Customer benefit*	*The idea is practicable and creates and adds value for the customer*
	User acceptance	*An existing demand is satisfied by the idea*
	Realizability	*The realization of the idea is economically feasible*
	Market size	*The expected demand of the target market justifies the idea's realization*
	Marketability	*The idea can be commercialized*
Quality	*Comprehensibility*	*The idea is written in an understandable way*
	Elaborateness	*The length of the description is adequate*
Business demands	*Risk*	*The risk of failure is compensated by the potential benefit for the company*
	Imitability	*The idea is sticky to the company's products and cannot be easily imitated by competitors*
Strategic fit	*Portfolio fit*	*The idea is expected to fit into the company's product portfolio*
	Development potential	*The idea is adaptable to new business requirements*

Table 5: Evaluation criteria for jury evaluations in Innovation Contests[17]

---

[17] Author's own table, referencing Ebner *et al.* (2009).

Alternatively, Riedl *et al.* (2010, p. 9) offer a set of 4 dimensions of idea quality evaluations ("novelty", "value", "feasibility", and "elaboration"), with 5-point rating scales from "low" to "high".

Regarding the design of the mechanisms for "rewarding" as the final step of the process (step F), this issue is of major interest in research into the economics perspective of Innovation Contests beside finding the optimum number and harvested efforts of solvers (Zheng *et al.*, 2011; Adamczyk *et al.*, 2012). The positive impact of rewards on participants highlights that "a reward was a significant determinant of a solver's performance" and "generally, the higher the reward, the higher the number of solutions" (Zheng *et al.*, 2011, p. 59) in a manner such that "larger prizes tend to stimulate higher performance" (Boudreau *et al.*, 2011, p. 3). The central question is how many prizes should be awarded. "In a fixed-price contest, the seeker announces a prespecified award with a fixed amount" and "instead of splitting a predetermined total award amount into two smaller awards, it is optimal to allocate the entire award to the best solution" (Terwiesch and Xu, 2008, p. 1534). Additionally, considering the expertise of the solvers, "a winner-takes-all award structure offers stronger incentive to solvers with high endowed expertise to exert effort because they are more likely to win the single award, whereas a multiple-prize award structure is more attractive to solvers with low endowed expertise because they have not much chance to win the first prize" (Terwiesch and Xu, 2008, p. 1534). Boudreau *et al.* (2011, p. 3) state that "single prizes are argued to be effective for homogeneous and risk-neutral individuals; multiple prizes are optimal when contestants have asymmetric ability and are risk averse". By highlighting the aspect of uncertainty for the contestants, Cohen *et al.* (2008, p. 434) mention that the "R&D contest is an example of a competition in which all contestants, including those who do not win any reward (prize), incur costs as a result of their efforts but only the winner gets the reward". As a solution, they invoke the effort-dependent rewards, described as rewards that "allow the designer of the contest to choose how the rewards, depend upon efforts" are distributed, which can lead to "substantial qualitative changes to the behavior of the contestants compared with constant-reward of contests" (Cohen *et al.*, 2008, p. 435).

Up to here, an overview of Innovation Contests, the central aspects of their design and selected research on additional aspects have been presented to provide a comprehensive view of Innovation Contests. Below, the theoretical perspectives for the topic of this thesis and subsequent investigations are detailed.

2.2   Theoretical perspectives: Organizational encouragement and support

Generally, the theory as the foundation of research activities is important, as highlighted, e.g., by Mayer and Sparrowe (2013): "Theories play an important role in management research. As management scholars, we utilize insights from many different theories as we engage in research to better understand many different aspects of management". In a similar manner, researchers state in a longer paragraph that "theories of the firm are conceptualizations and models of business enterprises which explain and predict their structure and behaviors. Although economists use the term 'theory of the firm' in its singular form, there is no single, multipurpose theory of the firm. Every theory of the firm is an abstraction of the real-world business enterprise which is designed to address a particular set of its characteristics and behaviors (Machlup, 1967). As a result, there are many theories of the firm which both compete in offering rival explanations of the same phenomena, and complement one another in explaining different phenomena" (Grant, 1996, p. 109).

Regarding the pertinent literature on Innovation Contests, Adamczyk et al. (2012) note that "one important shortcoming of the literature on Innovation Contests is the lack of theory. Up to now, there are no mainstream theories for exploring the research object of Innovation Contests" (Adamczyk et al., 2012, p. 355). Instead, scholars from various backgrounds (e.g., economics or information systems) have displayed an increasing research interest in the topic of Innovation Contests. Focusing on the number of theories to be considered in research activities, "many phenomena and research questions cannot be adequately addressed by drawing on a single theory" (Mayer and Sparrowe, 2013, p. 920).

Following the motivation for this thesis, the phenomenon and theoretical gap, the adaptation of two theories from a similar but differentiated research stream is pursued. More specifically, these theories are (i) the Componential Theory of Creativity and Innovation in Organizations and (ii) the Organizational Support Theory. Figure 8 depicts an overview of the theories that inform the topic of this thesis and therefore build its theoretical foundation. Both theories focus on the employer-employee relationship. First, the Componential Theory of Creativity and Innovation in Organizations is based in the research domain of creativity and innovation within organizations (Amabile et al., 1996; Amabile, 1988). At its core, the theory highlights the importance of employees' perceptions of their individual work environments as determinants for their creativity and organizational innovation (Amabile, 1997). The second organizational theory, the Organizational Support Theory, is based on the research domain of organizational (citizenship) behavior (Eisenberger *et al.*, 1990; Rhoades and Eisenberger, 2002). The theory assumes that employees with perceived organizational support (POS) feel obligated to help the organization through increased positive behavior, both in in-role activities and in extra-role activities (Aselage and Eisenberger, 2003).

Both theories highlight the existence of employees' perceptions for the explanation of their individual behavior. Whereas the Componential Theory of Creativity and Innovation in Organizations differentiates between *organizational encouragement* (organizational level) and *supervisory encouragement* (supervisory level), the Organizational Support Theory distinguishes *perceived organizational support* and *perceived supervisor support* (PSS), as explained in detail in the following.

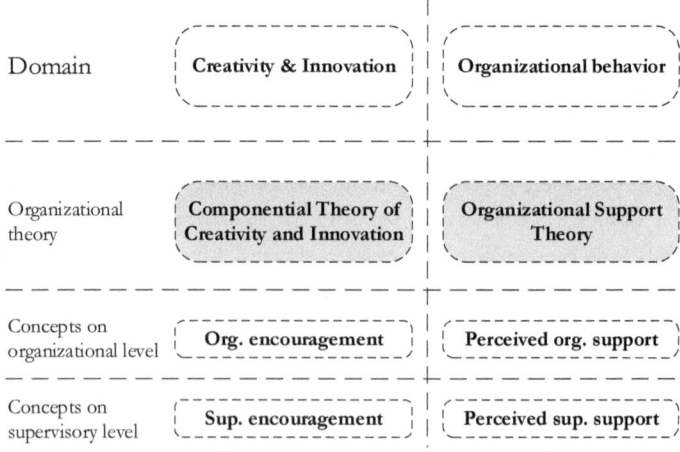

Figure 8: Theories informing central aspects of this thesis[18]

Below, a review of the emphasized organizational theories that inform this doctoral dissertation is presented.

2.2.1 The Componential Theory of Creativity and Innovation in Organizations

The following introduction of the Componential Theory of Creativity and Innovation in Organizations is separated into providing an overview and an outline of the history in section 2.2.1.1 and then by showing further details of the components of the theory in section 2.2.1.2.

---

[18] Author's own figure.

2.2.1.1 Overview and history of the theory

The Componential Theory of Creativity and Innovation in Organizations is designed and shaped by Teresa M. Amabile, Professor of Business Administration at Harvard Business School (Amabile *et al.*, 2014). It "has been tested and enlarged in a wide range of psychological studies over the past 25 years" (Amabile and Pillemer, 2012, p. 10). Teresa M. Amabile describes the theory in a manner in which the "aim of this theory is to adequately capture all of the major elements influencing creativity and innovation within organizations" (Amabile, 1997, p. 51). Central to her assumptions is the relationship between the individuals within organizations and their work environment when focusing on creative tasks: "The componential theory of creativity assumes that all humans with normal capacities are able to produce at least moderately creative work in some domain, some of the time—and the social environment (the work environment) can influence both the level and the frequency of creative behavior" (Amabile, 1997, p. 42). The theory highlights the importance of employees' perceptions of their daily work environment as important determinant of their work-related behavior (see Figure 9).

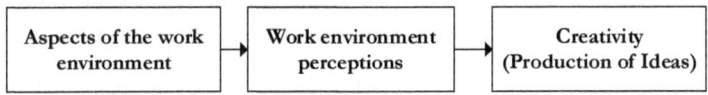

Figure 9: Social-environmental influence on employees[19]

As a basis for this analysis, diverse aspects of the work environment commonly influence employees' work environment perceptions, as stated in the literature: "Research on social environmental influences in organizations has uncovered aspects of the work environment [...]. Influences on work environment perceptions can arise at several different levels within an organization" (Amabile *et al.*, 1996, p. 1157). Furthermore, "several aspects are perceived as operating broadly across the organization" (Amabile *et al.*, 1996, p. 1159) and "it is the psychological meaning of environmental events that largely influences creative behavior" (Amabile *et al.*, 1996, p. 1158).

Work environment perceptions are described as employees' cognitive sense within organizations in which they are "forming and adjusting perceptions about the people you work with, the organization you are part of, the work you do" (Amabile and Kramer, 2007, p. 3). These highly individual perceptions are the foundation for increasing individuals' creativity and innovation within the firm, as highlighted in the following: "The central prediction of the theory is that elements of the work environment will impact individuals' crea-

---

[19] Author's own figure.

tivity [...]. The theory also proposes that the creativity produced by individuals and teams of individuals serves as a primary source for innovation within the organization" (Amabile, 1997, p. 52). Creativity is interpreted as "the production of novel and useful ideas" (Amabile *et al.*, 1996, p. 1155). Amabile and Kramer (2007, p. 1) illustrate the inner work life of employees as "the complex interplay between employees' deeply private perceptions of what's happening around them, the emotions they experience as a result of those perceptions, and their level of motivation to do good work". Creativity's influence on employees' motivation and performance in work activities is highlighted as "depending on what happens with these cognitive and emotional processes, motivation can shift, which, in turn, affects how people perform their work" (Amabile and Kramer, 2007, pp. 3 f.). They further suggest to "create conditions that enable people to get their work done, and you'll create positive emotions, enhance motivation, and boost performance to unprecedented levels" (Amabile and Kramer, 2007, p. 1). Amabile (1988, p. 126) defines individual creativity as follows: "Creativity is the production of novel and useful ideas by an individual or small group of individuals working together". Additionally, innovation within the organization is defined as follows: "Organizational innovation is the successful implementation of creative ideas within an organization" (Amabile, 1988, p. 126).

The foundation of the theory lies in several publications from the 1970s and 1980s (see Table 6 for details on selected publications). Below, a brief overview presents the milestones in the theory's development. In one of her first publications, Amabile (1979) proved the decrease in individuals' creativity when the expectation of an external evaluation is imposed upon students participating in an experiment involving artwork. Later, Amabile (1983) describes the components of individuals' creativity and their influence on the process steps of the response or idea generation by providing a componential framework for individual creativity. In addition, Amabile (1988) identifies a set of organizational factors that influence creativity and innovation in organizations. One year later, Amabile and Gryskiewicz (1989) published the "work environment inventory" (WEI) as the first instrument to assess the creative environment within firm boundaries. Additionally, the construct of work motivation and its specialties is illustrated later in Amabile (1993). Amabile *et al.* (1994) discuss the assessment of motivational orientations of individuals in organizations. In 1996, the "keys instrument" as a new approach for assessing the work environment perceptions was published. Based upon that approach, Amabile (1997) lists "expertise", "creative thinking", and "task motivation" as necessary components for individual creativity and provides eight scales for assessing the work environment for creativity (see later in this section). Later, a qualitative study investigated several leadership behaviors underlying a perceived leadership support (Amabile *et al.*, 2004). As noted above, the inner work life as a cyclic model that considers not only perceptions, emotions, and motivation related to work but also those factors' relationship to work performance is illustrated in Amabile and Kramer (2007). Amabile and Pillemer (2012) provide a retrospective on the social psychology of creativity and its development over the years. Finally, Amabile *et al.* (2014) offer insights into the design firm IDEO and their method of establishing a "culture of helping".

Author(s)	Main findings
Amabile (1979)	*Amabile (1979) conducted an experiment at Stanford University to investigate the influence of the expectation of an external evaluation on creativity. The results indicate that individuals in the group with a pending evaluation (especially those who have received nonspecific instructions) show decreased creativity compared to the control group.*
Amabile (1983)	*Amabile (1983) offers a componential conceptualization of individuals creativity by including the components of "domain-relevant skills", "creativity-relevant skills", and "task motivation". Additionally, the presented componential framework of individual creativity illustrates the influence of these components on the stages of the process for response or idea generation.*
Amabile (1988)	*This article describes an integrated model of individual creativity and organizational innovation. Within three interview studies, the author identified several personal qualities that promote (10 factors) or inhibit (5 factors) individual creativity and motivation (categorized into "domain-relevant skills", "creativity-relevant skills", and "task motivation"). Supplementary, qualities of the organizational setting for promoting (9 factors) or inhibiting (9 factors) are revealed. Finally, a model of organizational innovation with three components ("motivation to innovate", "resources in the task domain", and "skills in innovation management") is presented and connected to the individual creativity components.*
Amabile and Gryskiewicz (1989)	*The Work Environment Inventory as the first instrument for assessing the creative environment (with 135 items) is presented and compared to existing instruments (e.g., the Siegel Scale of Support of Innovation, The Creativity Audit, and the Innovation Climate Index).*

Amabile *(1993)*	*Details on work motivation seen as state or trait and motivational synergy among various types of extrinsic and intrinsic motivations are presented in this publication. Furthermore, the relationship and interactions between the two motivation categories are discussed.*
Amabile et al. (1994)	*This paper on motivational orientations discusses the theoretical background of the concept of motivation and its relationship to personality, attitudes and creativity. Additionally, the applicability of the work environment inventory for assessing the motivational orientations of employees is discussed.*
Amabile et al. (1996)	*The introduction and validation of "Keys: Assessing the climate for creativity" as a new instrument to assess the perceived work environment (stimulants and obstacles) is presented in this article and described as useful both for research and for practice. Furthermore, consideration of important quality criteria such as consistency, reliability, validity is proven.*
Amabile (1997)	*This article, titled "Motivating Creativity in Organizations", elaborated knowledge on expertise (domain-relevant skills), creative thinking (creativity-relevant skills), and task motivation as the three components of individual creativity. In addition, the work environment for creativity is detailed by presenting six stimulant scales (e.g., "organizational encouragement", "supervisory encouragement") and two obstacle scales ("organizational impediments", "workload pressure") that influence creativity and innovation.*
Amabile et al. (2004)	*The consideration of leadership behaviors and therefore perceived leader support in the assessment of a creative work environment is qualitatively investigated in this article. Therefore, different behaviors of leaders (positive and negative) related to perceived leader support are revealed (e.g., supporting, monitoring, recognizing, consulting, and clarifying roles and objectives).*
Amabile and Kramer (2007)	*The "inner work life" system as a complex interplay between employees' private perceptions, their emotions arising as the result of their perceptions, and influences on their work motivation are presented in this article. It highlights the mostly invisible interrelations of perceptions, emotions, and work motivation that are influenced by workday events and managerial actions and mainly affects employees' work performance.*
Amabile and Pillemer (2012)	*This paper provides a retrospective on the "social psychology of creativity". Therefore, inter alia, consensual assessment as enabler of social-psychological research in creativity, the intrinsic motivation principle of creativity, the Componential Theory of Creativity and Innovation in Organizations, and avenues for future research are presented.*

Amabile et al. (2014)	*By illustrating insights into the "culture of helping" within the design firm IDEO, Amabile and her colleagues highlight the importance of encouraging helping behavior within their firm as a major management challenge. They surveyed all the IDEO employees in one office and illustrated a dense network of assistance among all employees.*

Table 6: Selected research on the Componential Theory of Creativity and Innovation[20]

An overview of the application of the theory to various contexts is given by Amabile and Pillemer (2012). For a comparison of Amabile's and others' work, see McLean (2005).

Followed by the presented overview, further details on diverse components of the theory are presented next.

2.2.1.2 Components of the theory

As introduced above, the Componential Theory of Creativity and Innovation in Organizations distinguishes between one's own personality for individual creativity and the social or work environment that influences Creativity and Innovation in Organizations (Amabile, 1997, 1988). Therefore, the Componential Theory of Creativity and Innovation in Organizations consolidates the

- Componential Theory of **Individual** Creativity and Innovation in Organizations

and the

- Componential Theory of **Organizational** Creativity and Innovation in Organizations,

whereas "the organizational theory is built on the foundation of the Componential Theory of Individual Creativity and incorporates that theory" (Amabile, 1997, p. 52).

The theory is directed to the work environment and summarizes three organizational components ("Organizational motivation to innovate", "resources in the task domain", and "skills in innovation management"), which influence individuals' behavior related to creative tasks in an organizational context (Amabile, 1997, 1988). All the components summarize several single factors and activities by the managerial staff or the organization for encouraging employees to work on creative and innovative tasks (see Figure 10). For this interpretation, "creativity (or innovation) will be greatest in that area where all three

---

[20] Author's own table, in chronological order.

components overlap. This 'creativity intersection' defines the area of highest profitability for individual creativity or organizational innovation" (Amabile, 1988, p. 156). The organizational components are "intended to include all facets of the organization that might possible have an impact on the success of an innovation attempt" and "each of these components are essential for an organization to be innovative in its field" (Amabile, 1988, p. 153).

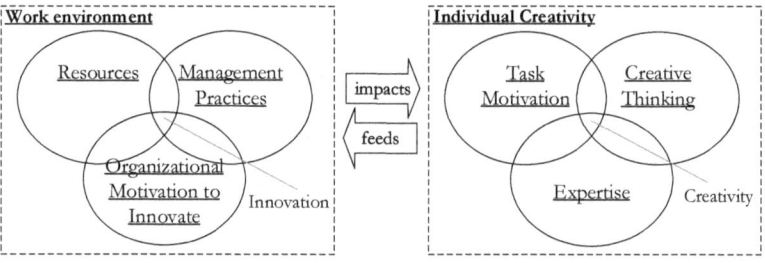

Figure 10: Individual and organizational components of the theory[21]

Domain-relevant skills ("expertise") and creativity-relevant skills ("creative thinking") can be affected by training, modeling, and experience afforded by the social environment. However, the most immediate and prevalent influence of the environment is exerted on the motivational component, as evidenced by empirical research on the "Intrinsic Motivation Principle of Creativity" (Amabile and Pillemer, 2012, p. 10).

The organizational component, "organizational motivation to innovate", focuses on the organization's basic orientation towards innovation. Support for that orientation must come from the C-Suite level and similarly, from middle management. Parts of this motivation to innovate are, e.g., a corporate vision and a clear overall goal of innovation, formulated general areas of innovation, and an orientation towards risk-taking, enthusiasm and an offensive lead towards the future (cf. Amabile, 1988, pp. 153 f.). The organizational component, "resources in the task domain", "includes everything the organization has available to aid work in the task domain [...] These resources include a wide array of elements" (Amabile, 1988, p. 154). For instance, people, funds, material resources, and market-research resources are important and "these various resources can be found in a variety of departments and divisions within organizations" (Amabile, 1988, p. 154). In summary, sufficient resources are one of the "prominent environmental promotors of creativity". The organizational component of "management practices" includes all management

---

[21] Author's own figure, referencing Amabile (1997, p. 53).

skills and styles supporting organizational innovation in several ways. Amabile summarizes, for instance, the following influencing factors (cf. Amabile, 1988, p. 155): Goal setting, participative and collaborative management, work assignments that consider both skills and interest, frequent feedback on work, and rewards and recognition for creative efforts.

In addition, the theory includes the components of individual creativity ("expertise", "creative-thinking skills", and "task motivation"). "According to the theory, these components combine a multiplicative fashion; none can be completely absent, if some level of creativity is to result" (Amabile and Pillemer, 2012, p. 10) On the individual level, "task motivation" "is responsible for initiating and sustaining the process; it determines whether the search for a solution will begin and whether it will continue, and it also determines some aspects of response generation" (Amabile, 1988, pp. 138 f.). In addition, "expertise" "includes all skills relevant to a general domain" (Amabile, 1988, p. 138). Finally, "creative thinking" influences "the way in which the search for responses will proceed" (Amabile, 1988, p. 139). Amabile highlights the impact of motivation: "In a sense, motivation is the most important of the three components, both for the individual and for the organization [...]. Domain-relevant skills and creativity-relevant skills determine what he or she is capable of doing, but the presence or absence of intrinsic task motivation will determine what that individual actually does" (Amabile, 1988, p. 156). Regarding individual creative performance, motivation "may be the most important component. No amount of skill in the domain or in methods of creative thinking can compensate for a lack of appropriate motivation to perform an activity" (Amabile, 1988, p. 133). In an exploratory study, Amabile (1988) identifies the qualities of individuals promoting or inhibiting creativity and innovation in organizations. In their study, being unmotivated was most often mentioned as an individual's negative quality. Being unmotivated is explained by a "lack of motivation for the work, not being challenged by the problem, having a pessimistic attitude toward the likely outcome; complacent, lazy" (Amabile, 1988, p. 129).

The individual components influence the process steps of idea generation in several ways, as illustrated in the Componential Model of Individual Creativity (see Figure 11). Amabile (1988, p. 138) explains that "this model describes the way in which an individual might assemble and use information in attempting to arrive at a solution, response, or product".

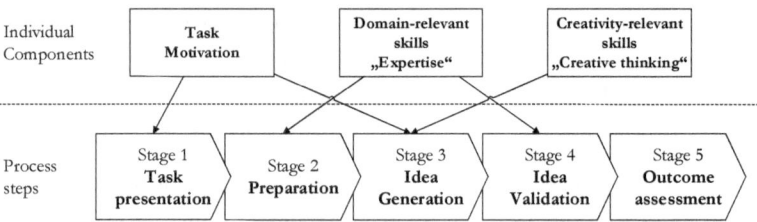

Figure 11: The componential model of individual creativity[22]

In the first stage, "task presentation", there "is the point at which the person becomes aware that there is an opportunity or a need to solve a problem or undertake a new task. Here, task motivation plays a prominent role because it determines whether and how the person will chose to engage" (Amabile and Pillemer, 2012, p. 10). For "preparation" (second stage), "domain-relevant skills play an important role as the person gathers information (and possibly learns new skills) in order to undertake the task" (Amabile and Pillemer, 2012, p. 10). In the third stage, "idea generation", "creativity-relevant skills and task motivation largely determine the outcome" (Amabile and Pillemer, 2012, p. 10). For "idea validation" (fourth) stage, "the person relies on domain-relevant skills to evaluate the novelty and usefulness of the candidate responses" (Amabile and Pillemer, 2012, p. 10). In the final stage, "outcome assessment", "the response is communicated and the outcome of the process is evaluated" (Amabile and Pillemer, 2012, p. 10).

In elaborating the theory, Amabile (2010, 1997) further differentiates the organizational components as essential parts of the Componential Theory of Creativity and Innovation in Organizations. Therefore, various perceptions that influence individuals' behavior involving creative tasks in an organizational context are defined and a distinction between positive and negative perceptions of the work environment with a (positive or negative) impact is made (Amabile, 1997). Positive perceptions are summarized as stimulant scales, and negative perceptions are summarized as obstacle scales.

For the organizational component of "organizational motivation to innovate", the theory differentiates between *"organizational encouragement"* (positive perception) and *"organizational impediments"* (negative perception). Regarding "resources", the scales *"sufficient resources"* (positive perception) and *"workload pressure"* (negative perception) are defined and distinguished. For the organizational component "management practice", the scales *"supervisory*

---

[22] Author's own figure, referencing Amabile (1988, p. 138).

encouragement", *"work group support"*, *"challenging work"*, and *"freedom"* are differentiated (all
positive perceptions). The description of the stimulant scales is presented in Table 7.

Stimulant scales	Scale descriptions
Organizational encouragement	*"An organizational culture that encourages creativity through the fair, constructive judgment of ideas, reward and recognition for creative work, mechanisms for developing new ideas, an active flow of ideas, and a shared vision of what the organization is trying to do."*
Supervisory encouragement	*"A supervisor who serves as a good work model, sets goals appropriately, supports the work group, values individual contributions, and shows confidence in the workgroup."*
Work group support	*"A diversely skilled work group in which people communicate well, are open to new ideas, constructively challenge each other's work, trust and help each other and feel committed to the work they are doing."*
Sufficient resources	*"Access to appropriate resources, including funds, materials, facilities, and information."*
Challenging work	*"A sense of having to work hard on challenging tasks and important projects."*
Freedom	*"Freedom in deciding what work to do or how to do it a sense of control over one's work."*

Table 7: Description of the creativity stimulant scales[23]

In addition to the positive perceptions of the work environment for creativity and innovation, the scales *"organizational impediments"* and *"workload pressure"* represent obstacle scales
that have a negative impact on individual creativity (see Table 8).

---

[23] Author's own table, referencing Amabile (1997) for direct quotations for scale descriptions.

Obstacle scales	Scale descriptions
Organizational impediments	*"An organizational culture that impedes creativity through internal political problems, harsh criticism of new ideas, destructive internal competition, an avoidance of risk, and an overemphasis on the status quo."*
Workload pressure	*"Extreme time pressures, unrealistic expectations for productivity, and distractions from creative work."*

Table 8: Description of creativity obstacle scales[24]

Finally, Amabile *et al.* (1996, pp. 1159 f.) mention several somewhat-broad groups of aspects that influence the above-introduced work environment perceptions. Table 9 offers the aspects that influence both a) *organizational encouragement* and b) *supervisory encouragement*.

Perceptions	Aspects influencing the perceptions
Organizational encouragement	*(1) Encouragement of risk taking and of idea generation* *(2) Fair, supportive evaluation of new ideas* *(3) Reward and recognition of creativity* *(4) Collaborative idea flow across organization, participative management, decision making*
Supervisory encouragement	*(1) Goal clarity* *(2) Open interactions between supervisors and subordinates* *(3) Supervisory support of a team's work and ideas*

Table 9: Aspects that influence selected work environment perceptions[25]

As illustrated in this section, Teresa M. Amabile focuses on work environment perceptions and their influence on individuals' creativity and organizational innovation. On a more general level, the level of organizational behavior, the Organizational Support Theory as the second theoretical lens informs the topic of this thesis, as presented below.

---

[24] Author's own table, referencing Amabile (1997).
[25] Author's own table, referencing Amabile *et al.* (1996, pp. 1159 f.).

2.2.2 Organizational Support Theory

2.2.2.1 Overview and history of the Theory

The Organizational Support Theory is primarily shaped by Robert Eisenberger and assumes *perceived organizational support* in a sense that "employees who receive highly valued resources (e.g., pay raises, developmental training opportunities) would feel obligated, based on the reciprocity norm, to help the organization reach its objectives through such behaviors as increased in-role and extra-role performance" (Aselage and Eisenberger, 2003, p. 492). Based on the concept of reciprocation (Eisenberger *et al.*, 1987; Eisenberger *et al.*, 2001), the authors state "that employees increase their efforts carried out on behalf of the organization to the degree that the organization is perceived to be willing and able to reciprocate with desirable impersonal and socioemotional resources" (Aselage and Eisenberger, 2003, p. 492). The impact of organizational support is assessed as important for a broad range of intra-organizational activities (Rhoades and Eisenberger, 2002).

The OST is based on the principle of social exchange (Shore and Wayne, 1993; Eisenberger *et al.*, 1986; Cropanzano, 2005) between the firm and its employees. Researchers highlighted the close connection between the social exchange theory (Blau, 1964) and the conceptualizations of *perceived organizational support*: "POS has long been conceptualized in SET terms" (Cropanzano, 2005, p. 883). Cropanzano (2005, p. 877) further highlighted that Eisenberger and his colleagues "were first to explore exchange ideology" and that "exchange ideology strengthens the relationship of POS and with felt obligation". At its core, "the social exchange view that employees' commitment to the organization is strongly influenced by their perceptions of the organization's commitment to them" (Eisenberger *et al.*, 1986, p. 500). Based upon and related to positive impact, "perceived organizational support is assumed to increase the employee's affective attachment to the organization and his or her expectancy that greater effort toward meeting organizational goals will be rewarded" (Eisenberger *et al.*, 1986, p. 500). Additionally, Shore and Wayne (1993, p. 775) state that "the social exchange framework that underlies POS suggests that these perceptions create feelings of obligation that serve to increase behaviors that support organizational goals".

The foundation of the OST is in the 1980s (see Table 10). A brief overview of the milestones of that theory is given below. In the beginning, Eisenberger *et al.* (1986) introduced the concept of *perceived organizational support* as the foundation of the OST. Subsequently, several positive relationships between employees' *perceived organizational support* and their job performance, affective commitment to the organization, and constructive innovation outcomes were revealed (Eisenberger *et al.*, 1990). Shore and Wayne (1993) proofed the positive influences of both *perceived organizational support* and affective commitment on employees' behavior. Additionally, the literature has verified a positive relationship between POS and employees' felt obligation, along with the indirect effect of felt obligation as a mediator of the relationships between POS and affective commitment and between POS and in-role performance (Eisenberger *et al.*, 2001). Rhoades *et al.* (2001) identifies diverse favorable work experiences ("organizational rewards", "procedural justice", and "supervisor support") as the antecedents of POS. Additionally, Eisenberger *et al.* (2002) verify both that *perceived supervisor support* leads to POS and that the relationship between PSS and employees' turnover is mediated by POS. A review of the pertinent literature on Perceived Organizational Support considering more than 70 studies is provided in Rhoades and Eisenberger (2002). In a meta-study, Eisenberger and Shanock (2003) discuss the relationships between rewards, intrinsic motivation, and creativity. In Shanock and Eisenberger (2006), several positive influences between POS and employees' POS and performance are proofed. The related concepts of a supervisor's organizational embodiment (Eisenberger *et al.*, 2010) and perceived organizational competence (Kim *et al.*, 2016) are introduced in separate articles.

Author(s)	Main findings
Eisenberger *et al.* (1986)	Eisenberger and his colleagues introduce perceived organizational support as a concept in which "employees form global beliefs about how organizations value their contributions and ensure care about individuals' well-being", formulating 36 statements to illustrate the construct.
Eisenberger *et al.* (1990)	This article provides the proof of a positive relationship between employees' POS and their individual job performance, affective commitment (involvement and attachment) and constructive innovation outcomes on behalf of the organization, e.g., giving suggestions for increasing the company's effectiveness (even in the absence of direct rewards or recognition).
Shore and Wayne (1993)	Shore and Wayne offer a contrasting view of affective commitment, continuance commitment and POS and their relevance as predictors of employee behavior. The results show that although POS and affective commitment both positively influence behavior, POS is assessed as the best predictor.

Eisenberger et al. (2001)	Eisenberger et al. investigate several relationships according to the reciprocation of POS. They prove that POS positively influences employees' felt obligation to the organization. Additionally, they verify that the relationships between POS and affective commitment and between POS and in-role performance are mediated by felt obligation. Additionally, a positive mood mediates the relationship of POS with affective commitment.
Rhoades et al. (2001)	This article mentions diverse favorable work experiences (organizational rewards, procedural justice, and supervisor support) that positively influence POS and affective commitment. Additionally, the revealed findings indicate that POS positively influences AC and negatively influences employee turnover.
Eisenberger et al. (2002)	Eisenberger and colleagues identify that perceived supervisor support leads to perceived organizational support. Furthermore, the relationship between PSS and POS is positively moderated by the perceived status of the supervisor. Finally, the mediating role of POS in the relationship between PSS and employees' turnover is verified.
Rhoades and Eisenberger (2002)	A meta-analysis (of more than 70 studies) identifies "fairness", "supervisor support", and "organizational rewards and job conditions" as antecedents of POS. Furthermore, e.g., "affective commitment", "job-related affect", "job involvement", and "performance" were identified as outcomes favorable to the organization.
Eisenberger and Shanock (2003)	In this article, a comparison of and the relationships between rewards, intrinsic motivation, and creativity are presented. The authors suggest a differentiation between performance for creative tasks (novel performance) and performance for standard tasks (conventional performance). They further state that on the one hand, rewards for novel performance might increase intrinsic motivation and creativity. On the other hand, rewards for conventional tasks decreases intrinsic motivation and creativity.

Shanock and Eisenberger (2006)	Supervisors with a high level of POS might have a higher obligation to help subordinates in their daily work. Consequently, a positive relationship between supervisors' perceived organizational support and subordinates' perceived supervisor support could be proofed in this study. Furthermore, subordinates' PSS leads to POS, in-role performance (INP), and extra-role performance (EXP) on a subordinate level.
Eisenberger *et al.* (2010)	The authors introduced supervisor's organizational embodiment (SOE) as the "extent of their supervisor's shared identity with the organization". They further proofed that SOE positively moderates the relationship between leader-member exchange (LMX) and affective organizational commitment. Additionally, a positive influence on both in-role performance and extra-role performance is verified.
Kim *et al.* (2016)	In this article, the authors introduced perceived organizational competence (POC) as "employees' perception concerning organization's ability to achieve its goals and objectives". They further prove that POC mediates the relationship between POS and AC and positively influences employees' extra-role performance.

Table 10: Selected research on the Organizational Support Theory[26]

2.2.2.2 Components of the theory

As highlighted above, the basis of the OST (Rhoades and Eisenberger, 2002) is built on a distinction between *perceived organizational support* and *perceived supervisor support* (Eisenberger *et al.*, 2002). Table 11 illustrates the corresponding descriptions. In terms of perceived support by the entire organization (POS), the company is assessed as an elementary counterpart of the employees and "employees showed a consistent pattern of agreement with various statements concerning the extent to which the organization appreciated their contributions and would treat them favorably or unfavorably in different circumstances" (Eisenberger *et al.*, 2002, p. 565). In terms of perceived support by the supervisor (PSS), direct superiors are seen as organizational representatives within the OST: "Employees are particularly aware that the directive, evaluative, and coaching functions of the supervisor are carried out on behalf of the organization, leading employees to generalize their views

---

[26] Author's own table, in chronological order

concerning the favorableness of their exchange relationship from supervisor to organization" (Eisenberger *et al.*, 2010, p. 2).

Concepts	Description
Perceived organizational support *(POS)*	*"Employees develop global beliefs concerning the extent to which the organization values their contributions and cares about their well-being"* (Eisenberger *et al.*, 2002, p. 565)
Perceived supervisor support *(PSS)*	*"Just as employees form global perceptions concerning their valuation by the organization, they develop general views concerning the degree to which supervisors value their contributions and care about their well-being" (Eisenberger et al., 2002, p. 565)*
Supervisor organizational embodiment *(SOE)*	*"Employees form a perception concerning the extent of their supervisor's shared identity with the organization: Supervisor's organizational embodiment, or SOE. The greater the SOE, the more the employee perceives that the supervisor shares the organization's characteristics"* (Eisenberger *et al.*, 2010, p. 2)
Perceived organizational competence *(POC)*	*"Employees' perception concerning the organization's ability to achieve its goals and objectives"* (Kim *et al.*, 2016, p. 1)

Table 11: Differentiated concepts within the Organizational Support Theory[27]

Later, the concepts of "supervisor organizational embodiment" and "perceived organizational competence" were published. A detailed introduction of SOE and POC is not provided here; instead, a reference to the original publications is made (see above).

Regarding the implications or consequences of POS, "in-role performance", "extra-role performance" and "organizational citizenship behavior" (OCB) are highlighted as behavioral outcomes (Kurtessis *et al.*, 2015; Rhoades and Eisenberger, 2002), as illustrated in Table 12[28]. In addition to an increase in the performance of standard job activities (so-called "in-role performance", (Rhoades *et al.*, 2001)), POS positively influences employees' "extra-role performance". This includes behavior such as helping other employees or "offering constructive suggestions, and gaining knowledge and skills that are beneficial to the organization" (Rhoades and Eisenberger, 2002, p. 702).

---

[27] Author's own table.
[28] See Rhoades and Eisenberger (2002) for a comprehensive review of the antecedents and effects of POS.

Performance dimen-	Description
In-role performance (INP)	*The employee, e.g., "meets formal performance requirements of the job"; "fulfills responsibilities specified in job description"; and "performs tasks that are expected of him or her"* (Eisenberger et al., 2001, p. 45)
Extra-role performance (EXP)	*"Actions favorable to the organization that go beyond assigned responsibilities"* (Rhoades and Eisenberger, 2002, p. 702)
Organizational citizenship behavior (OCB)	*"Every factory, office, or bureau depends daily on a myriad of acts of cooperation, helpfulness, suggestions, gestures of goodwill, altruism, and other instances of what we might call citizenship behavior"* (Smith et al., 1983, p. 653)

Table 12: Performance dimensions distinguished in the Organizational Support Theory[29]

Supplementary, "organizational citizenship behavior" is assessed as "a category of performance" that "goes beyond formal role requirements" (Smith *et al.*, 1983, pp. 653 f.). Representing the degree of altruism and helping persons within an organization, "substantively, citizenship behaviors are important because they lubricate the social machinery of the organization" (Smith *et al.*, 1983, pp. 653 f.).

In sum, the OST is based on the principle of reciprocation and postulates that employees who perceive a high level of organizational support (*POS* and *PSS*) are more likely to help their co-workers and/or the organization. This help includes diverse performance dimensions, including employees' INP, EXP, and OCB.

In addition to both of the above-introduced theories that build the foundation for this thesis, the Rubicon model of action phases offers interesting insights regarding the explanation of individuals' actions and behavior, as explained in the below, and therefore is highlighted here.

2.2.3 The Rubicon model of Action Phases

From a theoretical perspective, individuals' actions and behavior could be explained by the Rubicon model of Action Phases (Heckhausen and Gollwitzer, 1986, 1987). The model posits four distinct phases (deliberating, planning, acting, evaluating) during motivational- and volitional-oriented behavior that regularly begins with the awakening of a person's wishes. The model considers (1) a pre-decisional phase, (2) a post-decisional but

---

[29] Author's own table.

pre-actional phase, (3) an actional phase, and (4) a post-actional phase (see Figure 12). The pre-decisional phase is shaped in a manner that is mainly motivational. Intrinsic motivation is defined as personal engagement in work because it is interesting or satisfying, whereas extrinsic motivation indicates the motivation to work for "something apart from and external to the work itself" (Zheng *et al.*, 2011, p. 61). Both intrinsic and extrinsic motivation influences the decision to participate in online contests (Leimeister *et al.*, 2009; Frey *et al.*, 2011). The pre-actional phase is shaped as primarily volitional.

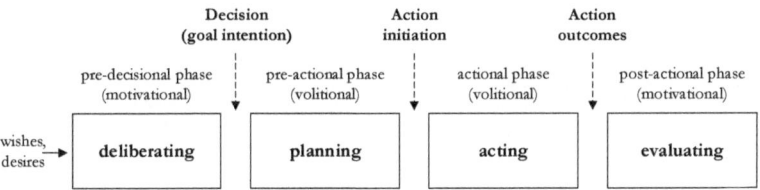

Figure 12: Action phases of the Rubicon-model

Below, a brief summary of the action phases of the Rubicon model is given (see Gollwitzer *et al.*, 1990 for more details): The "pre-decisional action phase" is characterized by wishes and deliberating. For various reasons, individuals develop wishes and desires. People often must deliberate and decide among several wishes that they want to pursue based on criteria such as feasibility, desirability and expectations. Important factors are the expected value and short- and long-term consequences. At the end of the pre-decisional action phase, the decision is made ("crossing the Rubicon") and wishes are transformed into a goal intention. During the "pre-actional phase", questions such as when, where, how and how long to act are addressed during the planning phase. Committing oneself to several actions forms the behavioral intention. Researchers also note the psychological empowerment of employees that results from having their activities supported within their personal work environment, e.g., creative process engagement (Zhang and Bartol, 2010; Spreitzer, 1995). Action initiation builds the transition to the actional phase and depends on the volitional strength and the favorability of any opportunities to act. During the "actional phase", behavior is characterized by actions toward goal achievement that are directed by the mental representation of the goal. Volitional strength influences the individual's effort exertion. Finally, during the "post-actional phase", individuals evaluate whether the achieved goal matches their expectations.

3 STUDY A:

The work environment and participation in Innovation Contests

In the chapter 3, the conceptualization and execution of the first study of work environment perceptions and their impact on participation in Innovation Contests is presented. For this empirical investigation of the influence of work environment perceptions as a determinant of employees' *motivation, affective organizational commitment*, and *participation intention*, the use of a web survey and multi-variate data analysis methods (exploratory and confirmatory factor analysis, structural equation modeling, and moderation and mediation analysis) are presented. For this purpose, the detailed aim and study design are described in line with the introduction in chapter 1 (see section 3.1). Chapter 3 then proceeds as follows. The presentation of the conceptual model, relevant constructs, their relationships and the formulated hypotheses are provided (section 3.2). Next, further details of the data-collection procedures (section 3.3) and the data-analysis procedures (section 3.4) are given. Chapter 3 ends with the presentation and discussion of the empirical findings (section 3.5) and the summary of study A (section 3.6).

3.1 Aim and study design

As stated in the introduction to this doctoral dissertation, Study A

"… empirically investigates different work environment perceptions (both positive and negative, and including organizational support) and those perceptions' influence on employees' *motivation*, on *affective organizational commitment* and on *participation intention*" (Section 1.5.2, "Research design").

The corresponding research question 1 (see section 1.2, "Derivation of the research questions") reads as follows: "What is the influence of different work environment perceptions on employees' affective organizational commitment and on their motivation and intention to participate in firm internal Innovation Contests?"

In this study, an empirical investigation is executed in cooperation with employees from a large DAX 30 company in Germany both to answer the research question and to investigate the causalities and hypotheses using a survey as the non-experimental design for data collection and statistical analytics (see Table 13).

Properties	Criteria
Nature of study	Explanatory, hypothesis testing
Study design	Quantitative, causal relationships
Research method	Online survey and statistical analytics

Table 13: Overview of the study design for Study A[30]

The conceptualization and preliminary findings of this study were published[31] at the European Conference on Information Systems (Hoeber *et al.*, 2016).

The intended theoretical relevance and contributions of this dissertation are highlighted in the first chapter of the thesis, which states, "This dissertation sheds light on work environment perceptions and organizational support as an important determinant and critical success factor for Innovation Contests" (Section 1.3, "Relevance and contributions"). With respect to this study's theoretical foundation and proposed conceptual model, its challenge is to "conceptually integrating the literature on organizational encouragement and support (the "Componential Theory of Creativity and Innovation in Organizations" and the "Organizational Support Theory" (OST)), motivation and commitment theory, and Innovation Contests" (Section 1.3, "Relevance and contributions") and its application to the specific context of firm internal Innovation Contests. Therefore, the main effort lies in applying the theories and investigating several effects not only on employees' *motivation* and *affective organizational commitment* as central personal attitudes but also on their *participation intention* as an important outcome variable. With this study, a better understanding of the influence of several positive and negative work environment perceptions in the context of Innovation Contests, hosted by large, often multi-divisional firms, is the goal.

To collect data, the call to participate in this study was sent by email to approximately 750 employees of a large company in Germany's telecommunications industry. The data collection is done exclusively within one of the firm's subunits, the "product house" for the group, which is responsible for further development of the firm's entire product and service portfolio. Prior to the field period, a pre-test of the questionnaire was conducted with

---

[30] Author's own table.
[31] As the first author, Bjoern Hoeber is entirely responsible for the content of the article.

colleagues and target respondents to improve and refine the survey instrument (cf. Presser *et al.*, 2004).

For data analysis, a two-step approach is chosen following Anderson and Gerbing (1988). First, to test the measurement model, exploratory and confirmatory factor analyses (EFA, CFA) are applied to assess the dimensionality, reliability and validity of the measurements and constructs. Second, testing the structural model is conducted both to evaluate the model fit and to test the links between the constructs by analyzing the sign, magnitude and statistical significance of the path coefficients. Additionally, moderation and mediation analysis with central variables are executed to further improve knowledge of the relationships between the investigated constructs.

To conclude, a summary and discussion of the results is provided. In the next section, a detailed overview of the proposed research model is presented.

3.2 Overview of the conceptual model

After first presenting the proposed conceptual model (section 3.2.1), the constructs, their relationships and the derivation of the hypotheses are noted in the latter part of this section (section 3.2.2).

3.2.1 The proposed conceptual model

This study and the derived conceptual model focus on (i) work environment perceptions (both positive and negative), (ii) employees' *motivation* and *affective organizational commitment* and (iii) *participation intention* in the context of corporate Innovation Contests (see Figure 13).

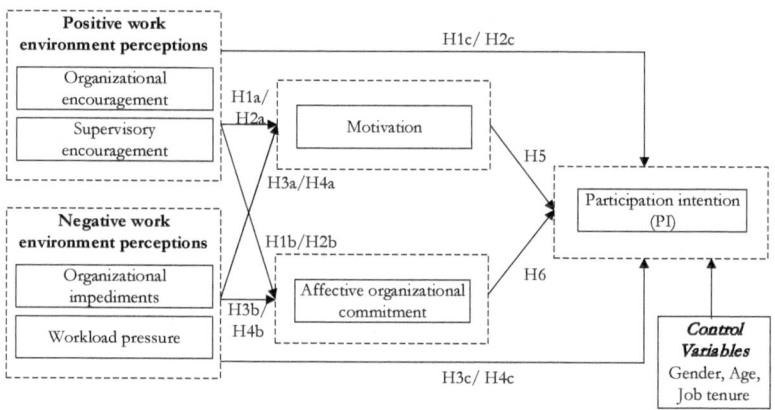

Figure 13: The conceptual model for Study A[32]

The proposed conceptual model includes work environment perceptions as both positive (*organizational encouragement, supervisory encouragement*) and negative (*organizational impediments, workload pressure*) determinants (exogenous variables), in line with Amabile *et al.* (1996). Next, there is a focus on *motivation* and *affective organizational commitment* (as intervening variables) because "both have been described as energizing forces with implications for behavior" (Meyer *et al.*, 2004, p. 994). The relationship between employees' *motivation* and their *affective organizational commitment* is investigated by Meyer *et al.* (2004), with the result that both concepts are "distinguishable, albeit related" (Meyer *et al.*, 2004, p. 991). To measure individuals' willingness to participate in Innovation Contests, *participation intention* (cf. Alexandris *et al.*, 2007; Zheng *et al.*, 2011) is considered (endogenous variable). All the items were adapted and tailored to the context of this dissertation.

Below, a brief introduction of the different variables is provided.

First, a call to consider the social environment within the firm when investigating employees' individual motivation and creativity has been made by Teresa M. Amabile (1997, p. 40):

> *"Although part of intrinsic motivation depends on personality, my students, colleagues, and I have discovered in 20 years of research that a person's social environment can have a significant effect on that person's level of intrinsic motivation at any point in time; the level of intrinsic motivation can, in turn, have a significant effect on that person's creativity."*

---

[32] Author's own figure.

Therefore, this study focuses on four **work environment perceptions** that have a striking influence on creative behavior in organizations (see Amabile, 1997, pp. 49–50) and explain the influence on employees' *motivation* and *affective organizational commitment.* "Clearly, although all aspects of the work environment may exert influence, some appear to carry more weight in the differentiation" (Amabile, 1997, p. 49). These reasons are grouped into (i) positive and (ii) negative work environment perceptions (based on the distinction between stimulant and obstacle dimensions, see Amabile *et al.* (1996); Amabile (1997)). All these reasons induce a higher (for positive perceptions) or lower (for negative perceptions) participation intention associated with a supportive work environment. Regarding positive work environment perceptions, *organizational encouragement* and *supervisory encouragement* are both important. As negative perceptions towards the work environment, *organizational impediments* (OI) and *workload pressure* (WP) are important dimensions identified by Amabile *et al.* (1996) and describe why employees are hampered at participation in creative work: "Two scales focus on environmental obstacles to creativity—factors that should be negatively related to creative work outcomes—including organizational impediments and excessive workload pressure" (Amabile, 1997, p. 46).

Employees' work *motivation* has been investigated not only in several contexts of organizational studies (Pinder, 2014) but also in a few investigations within the area of contests for innovative tasks and idea creation (Zheng *et al.*, 2011; Leimeister *et al.*, 2009; Frey *et al.*, 2011). This study considers the *motivation* construct because *motivation* is the "energizing force—it is what induces action in employees" (Meyer *et al.*, 2004, p. 992). Furthermore, *motivation* is important for creative tasks, as it is in Innovation Contests (Adamczyk *et al.*, 2012), because "motivating people to perform at their peak is especially vital in creative work. An employee uninspired to wrap her mind around a problem is unlikely to come up with a novel solution" (Amabile and Khaire, 2008, p. 106). Working on creative tasks requires expertise and skills, but "although a person's development of expertise and practice of creative thinking skills can be influenced to some extent by the social environment, the strongest and most direct influence of the environment is probably on motivation" (Amabile, 1997, p. 44).

There are many forms and definitions of work motivation; for a comparison of the various forms of motivation (i.e., work motivation, intrinsic motivation, extrinsic motivation) and for the relationships and effects among these various forms, see Amabile (1993). A differentiation between intrinsic and extrinsic motivational orientations is also emphasized by Amabile *et al.* (1994, p. 950): "The 'labor of love aspect' driving human behavior is what psychologists have, for several decades, called intrinsic motivation: The motivation to engage in work primarily for its own sake, because the work itself is interesting, engaging, or in some way satisfying". However, "the contrasting concern [...] fits the definition of extrinsic motivation: The motivation to work primarily in response to something apart from the work itself, such as reward or recognition or the dictates of other people" (Amabile *et al.*, 1994, p. 950).

The "intrinsic motivation principle of creativity" (Amabile and Pillemer, 2012; Amabile, 1997, 1988) claims that "the intrinsically motivated state is conducive to creativity, whereas the extrinsically motivated state is detrimental" (Amabile and Pillemer, 2012, p. 7). Furthermore, researchers have stated that the Componential Theory of Creativity and Innovation in Organizations expanded the "research into the effects of social-environmental factors on intrinsic motivation" (Amabile and Pillemer, 2012, p. 8). In this thesis, the focus is on firm internal Innovation Contests that aim both to further develop a firm's products and services and to improve work-related conditions (e.g., processes, procedures). Therefore, the focus is portrayed more in an intrinsic motivational orientation or "intrinsic task interest" (Eisenberger and Shanock, 2003, p. 121) in which "motivation arises from the individual's perceived value of engaging in the task itself (e.g., finding it interesting, enjoyable, satisfying, or positively challenging)" (Amabile and Pillemer, 2012, p. 7). The motivation variable in this study is anchored to the context of Innovation Contests, meaning a targeted measuring of the strength of employees' motivation for participation in firm internal Innovation Contests.

Employees' *affective organizational commitment* is an important individual attitude described as a "binding of the individual to an organization" (Meyer and Allen, 1991, p. 73). Affective organizational commitment is defined as "an affective or emotional attachment to the organization such that the strongly committed individuals identifies with, is involved in, and enjoys membership in, the organization" (Allen and Meyer, 1990, p. 2). Commitment is highlighted as "a central focus of research in organizational behavior [...] to understand how the psychological bonds that arise between employees and organizations influence workplace behaviors" (Bateman et al., 2011, p. 842). Additionally, *affective organizational commitment* is assessed to be "important in predicting nonrole behaviors" (Shore and Wayne, 1993, p. 775) and is accepted as a "powerful predictor" (Bateman et al., 2011, p. 842).

Focusing on online communities, Bateman et al. (2011) analyze the difference between commitment in general and commitment in the context of online communities with the result that the effects are analogous, not identical. Here, *affective organizational commitment* has a positive influence on individuals' posting behavior: Individuals "want to be part of the conversation" and "find their association with it to be emotionally fulfilling" (Bateman et al., 2011, pp. 843 ff.). Additionally, the literature highlights a significant positive relationship between individuals' identification with the community (expressed as "affective social identity" or "social influence") and their *participation intention*. *Affective commitment* leads individuals "to want to help others who are part of their community by engaging in conversation with them" and *affective commitment* "helps ensure the long-term success of the community by making it more likely that questions will receive responses" (Bateman et al., 2011, pp. 849–850).

Until now, no studies investigating the relationship between *affective organizational commitment* and *participation* in Innovation Contests have been found.

Comparing the concepts of *motivation* and *affective organizational commitment*, the following contrast both illustrates and strengthens the choice for the integration into the conceptual model that represents and addresses various influences. On the one hand, the concept of *motivation* indicates that individuals' behavior in doing (or not doing) something is influenced by a range of personal motives, such as learning, direct compensation, self-marketing, and social motives (cf. Leimeister *et al.*, 2009). Additionally, the intrinsic motivation principle of creativity highlights the importance of intrinsic task *motivation* for creative work (Amabile, 1997). On the other hand, the concept of *affective organizational commitment* focuses on employees' behavior in terms of their will to help the organization out of a sense of duty, conscientiousness, or obligation towards the employer (Rhoades and Eisenberger, 2002).

In addition, a further facet regarding the differentiation in the constructs' anchors can be highlighted with respect to this study's conceptualization. The anchors of *motivation* and *participation intention* are both explicitly located in the scenario of firm internal Innovation Contests. In contrast, the *affective organizational commitment* construct is "anchored in an organizational frame of reference" (Pierce *et al.*, 1989, p. 624) and presents a more clearly distal or generic concept compared to motivation and its connection to participation intention. Here, this study intends to offer a comparison in the above described manner.

Finally, this study analyses the impact of the work environment perceptions on employees' *motivation, affective organizational commitment* on their *participation intention* as the main outcome variable in the context of corporate Innovation Contests. The working definition of *participation intention* is derived from Ajzen (1991, p. 181): "Intentions are assumed to capture the motivational factors that influence a behavior; they are indications of how hard people are willing to try, of how much of an effort they are planning to exert, in order to perform the behavior. As a general rule, the stronger the intention to engage in a behavior, the more likely should be its performance." Participation intention is a suitable choice for this research because it is assessed and proven as a reliable predictor of real participation (Kim *et al.*, 2008; Ajzen, 1991; Zhou, 2011). It is also argued by Zheng *et al.* (2011, p. 61) that "the theory of planned behavior (1) posits that an individual actual behavior can be predicted by the intention to perform the behavior".

In sum, the conceptual model consists of four independent variables that represent several work environment perceptions (*organizational encouragement, supervisory encouragement, organizational impediments, and workload pressure*), two mediator variables (*motivation and affective organizational commitment*) and one outcome variable (*participation intention*). Next, the assumed relationships between the latent variables are presented.

3.2.2   Relationships between latent variables and hypotheses development

A description of the relationships between the latent variables in the context of this study and a derivation of the research hypothesis is given below. Therefore, the relationships are detailed and discussed in adherence with the direction and their appearance in the conceptual model, starting with the work environment perceptions on the left hand side representing the reasons as independent variables. In addition to the perceptions that offer general theoretical access to the characterized relationships, additional relevant literature is highlighted and summarized for a comprehensive representation of the influences below.

3.2.2.1   The direct influences of organizational encouragement

The positive influence of *organizational encouragement* both generally and on (a) employees' motivation, (b) employees' affective organizational commitment, and (c) participation intention in firm internal Innovation Contests (Hypotheses H1a, H1b, H1c)

Organizational encouragement is seen as a type of organizational culture that encourages mechanisms for developing novel ideas, active idea flows, and fair judgments of ideas within the organization (Amabile *et al.*, 1996). It is assumed to be a decisive factor in employees' behavior in Innovation Contest contexts. Following Amabile *et al.* (1996, p. 1159 f.), *organizational encouragement* includes the following four aspects: (i) encouragement of risk taking and idea generation, (ii) a fair, supportive evaluation of new ideas, (iii) reward and recognition of creativity, and (iv) collaborative idea flow, participative management and decision making. As expressed by Amabile (1997, p. 55), "organizations must demonstrate a strong orientation toward innovation, which is clearly communicated and enacted, from the highest levels of management, throughout the organization". Expressing the impact of an encouragement by the organization, for example, Eisenberger et al. (1990) have confirmed that perceived organizational support is related to "innovation on behalf of the organisation" (Eisenberger et al., 1990, p. 51) and that "perceived support might be associated with constructive innovation [...] without the anticipation of direct reward or personal recognition" (Eisenberger et al., 1990, p. 54).

The psychological process underlying individuals' reactions to POS could be separated into different phases (cf. Rhoades and Eisenberger, 2002). First, when a certain degree of organizational support is perceived by employees, it produces "a felt obligation to care about the organization's welfare and to help the organization reach its objectives" (Rhoades and Eisenberger, 2002, p. 699). This leads to an incorporation of organizational membership into employees' social identity and subsequently to a strengthened belief that the organization recognizes and rewards increased performance. Finally, this process shows positive outcomes for both sides—the employees (e.g., satisfaction) and the firm (e.g., increased commitment and performance). Below, the relationships between *organizational encouragement* and the corresponding dependent variables of *motivation, affective organizational commitment,* and *participation intention* are detailed.

Regarding employees' *motivation* and the impact of *organizational encouragement,* it is highlighted that a "person's social environment can have a significant effect on that person's level of intrinsic motivation at any point in time" (Amabile, 1997, p. 40). The influence of *organizational encouragement* on *motivation* is additionally expressed by the following quote: "When people form negative perceptions—of their manager, organization, coworkers, work, or themselves—they feel frustrated and unhappy. Motivation shrivels. [...] But when employees form positive perceptions, the cycle turns from vicious to virtuous" (Amabile and Kramer, 2007, p. 1). Amabile *et al.* (1996, p. 1159 f.) clarified the importance of *organizational encouragement* for employees' motivation in idea generation, idea evaluation, idea flow and feedback, all which are central steps in Innovation Contests. For idea generation, "psychological research on creativity has demonstrated that people are more likely to produce unusual, useful ideas if they are given licence to do so" and for idea evaluation, studies have "demonstrated that supportive, informative evaluation can enhance the intrinsically motivated state that is most conductive to creativity". Regarding the flow of various ideas within organizations, "research has shown that the probability of creative idea generation increases as exposure to other potentially relevant ideas increases" (Amabile *et al.*, 1996, p. 1160). Finally, regarding the feedback dimension of *organizational encour-*

agement, the literature has indicated that "when employees have high autonomy, receive feedback about their performance [...] they may experience feelings of happiness, and hence intrinsic motivation to keep performing well" (Houkes *et al.*, 2003, p. 428). Considering the effects of *organizational encouragement* highlighted above, a similar positive effect on employees' *motivation* is expected in the Innovation Contest context. Therefore, the first hypothesis reads as follows:

Hypothesis 1a: *A higher level of perceived organizational encouragement is positively related to employees' motivation for participating in Innovation Contests.*

Regarding employees' *affective organizational commitment, organizational encouragement* is expressed as a decisive factor influencing employees' *affective organizational commitment*. Perceived support by the organization might express stronger feelings of employees' affiliation and loyalty (Eisenberger et al., 1990). Rhoades *et al.* (2001) and Shore and Wayne (1993) have found that POS leads to higher *affective organizational commitment* on employees' side: "POS would enhance AC by producing a felt obligation to care about the organization's welfare and by the incorporation of organizational membership and role status into social identity" (Rhoades *et al.*, 2001, p. 826). Here, the reciprocity principle is essential (Rhoades and Eisenberger, 2002); the enhancement of *affective organizational commitment* is explained as the consequence of "fulfilling such socioemotional needs as affiliation and emotional support. [...] Such need fulfillment produces a strong sense of belonging to the organization, involving the incorporation of employees' membership and role status into their social identity" (Rhoades and Eisenberger, 2002, p. 701). Additionally, Meyer *et al.* (2002, p. 38) report that to increase *affective organizational commitment*, firms must "treat employees fairly and provide strong leadership. [...] We also found that affective commitment correlates strongly with the various forms of organizational justice (i.e., distributive, procedural, and interactional)". The element of organizational justice is explicitly represented in the *organizational encouragement* construct as fair judgments of ideas (Amabile *et al.*, 1996). Because employees' *affective organizational commitment* is a general attitude with an organizational context of reference (cf. Shore and Wayne, 1993) and is not specific to Innovation Contests, a similar positive influence is assumed in this study. Therefore,

Hypothesis 1b: *A higher level of perceived organizational encouragement is positively related to employees' affective organizational commitment.*

Regarding employees' *participation intention*, a positive effect attributable to *organizational encouragement* is assumed based on the general relationship between POS and performance. POS "should increase performance of standard job activities and actions favorable to the organization that go beyond assigned responsibilities" (Rhoades and Eisenberger, 2002, p. 702). Eisenberger *et al.* (1990, p. 51) indicate that organizational support is related to employees' willingness to help generate "innovation on behalf of the organization"; moreover, "perceived support was also positively related to employee innovation. The constructiveness of anonymous, voluntary suggestions for improving the organization was greater by those perceiving that the organization valued their contribution and cared about their well-being. Such innovative suggestions are important to the organization's growth and success" (Eisenberger *et al.*, 1990, p. 57). It is exactly these types of voluntary and innovative suggestions that represent the contributions that Innovation Contests attempt to collect with their possibilities.

Typically, employee participation in Innovation Contests is voluntary participation (Adamczyk *et al.*, 2012; Hutter *et al.*, 2011) that is not included in their contractual agreements and that is based on people's "willingness to freely reveal their knowledge and expertise and openly work together" (Hutter *et al.*, 2011, p. 8). Here, Chen *et al.* (2009) note that "the relation between POS and extra-role performance, involving activities that aid the organization but are not explicitly required of employees, was stronger than the relation between POS and performance of standard job activities" (Chen *et al.*, 2009, p. 120). They explain extra-role behavior as, e.g., "offering constructive suggestions, and gaining knowledge and skills that are beneficial to the organization" (Chen *et al.*, 2009, p. 120). In this context, behavior in extra-roles is often mentioned together with the concept of organizational citizenship behavior. For example, Shore and Wayne (1993, p. 775) highlight that "Organizational citizenship behavior (OCB) is extra-role behavior that is generally not considered a required duty of the job or part of a traditional job description".

Additionally, the aspect of fairness should be highlighted when considering individual decisions to participate in firm innovation, as highlighted by Franke et al. (2013). They argue that various forms of perceived fairness are significant to individuals' willingness to contribute. Here, the manner in which organizations design and execute the underlying process (so-called procedural fairness) and the manner in which firms define the contribution of the outcomes (so-called distributive fairness) are highlighted and proofed to influence users' participation intention.

Although the scientific literature has paid less attention to the linkages of *organizational encouragement* and support in the area of corporate Innovation Contests, similar effects are assumed. This study follows the above-described positive effects of an *organizational en-*

couragement to participate in Innovation Contests, with *participation intention* seen as extra-role activity or performance. Taken together, hypothesis H1c is as follows:

Hypothesis 1c: *A higher level of perceived organizational encouragement is positively related to employees' intention to participate in Innovation Contests.*

3.2.2.2 The direct influences of supervisory encouragement
The positive influence of *supervisory encouragement* both generally and on (a) employees' motivation, (b) employees' *affective organizational commitment*, and (c) *participation intention in firm internal Innovation Contests* (Hypotheses H2a, H2b, and H2c)

Supervisory encouragement is seen as supporting individual contributions; support of the work group might also have a positive influence that is assumed a determinant of employees' behavior. In addition to perceptions of encouragement and support by the entire organization (as highlighted in the former section), the second influence is sourced in the area of encouragement by direct supervisors, as indicated by the following quote from the organizational behavior literature: "Because supervisors act as agents of the organization, having responsibility for directing and evaluating subordinates' performance, employees view their supervisor's favorable or unfavorable orientation toward them as indicative of the organization's support" (Rhoades and Eisenberger, 2002, p. 700). As highlighted in the Componential Theory of Creativity and Innovation in Organizations, *supervisory encouragement* is defined as direct supervisors' support function related to creative and innovation-oriented tasks; *inter alia*, for facilitating open interactions, clarification of goals (Amabile et al., 1996): "A slowly expanding body of literature over the past 30 years has documented the importance of perceived leader support for subordinate creativity" (Amabile *et al.*, 2004, p. 7).

Below, the relationships between *supervisory encouragement* and the corresponding dependent variables of *motivation, affective organizational commitment*, and *participation intention* are detailed.

Regarding employees' *motivation* and focusing on the role of direct supervisors (*supervisory encouragement*), according to Amabile *et al.* (1996), important encouragement includes clarifying goals and tasks, fostering open interactions and supporting both work and ideas: "Under these circumstances, people are less likely to experience the fear of negative criticism that can undermine the intrinsic motivation necessary for creativity" (Amabile *et al.*,

1996, p. 1160). Following Amabile *et al.* (2004, p. 6), the Componential Theory of Creativity and Innovation in Organizations "proposes perceived leader support (often termed as *'supervisory encouragement'*) as the feature that is under the most direct control of the immediate supervisor. [...] The support provided by immediate supervisors exerts an influence on subordinates' creativity through direct help with the project, the development of subordinate expertise, and the enhancement of subordinate intrinsic motivation".

In a similar vein, the literature often mentions the direct and positive effect of managerial behaviors on motivation; Amabile *et al.* (2004) investigate several leader behaviors as antecedents of PSS, announcing that a broad range of behavioral actions are essentially as follows: "A good leader can do much to challenge and inspire creative work in progress. [...] The wrong managerial behaviors, or simply careless neglect, can be tremendously demotivating" (Amabile and Khaire, 2008, p. 107). Amabile *et al.* (2004) describe a demotivating situation in which the supervisor seldom asks team members' "input into decisions. This lack of consultation not only appeared to undermine subordinates' motivation to give the project their best efforts, but it also likely deprived the project of fresh perspectives that could have saved it" (Amabile *et al.*, 2004, p. 24). Another quotation indicates that "it may stun you, if you are a manager, to learn what power you hold. Your behavior as a manager dramatically shapes your employees' inner work lives" (Amabile and Kramer, 2007, p. 2). Zhou and Shalley (2003) claim that developmental feedback by the immediate superior increases intrinsic motivation and creativity. To recap, *supervisory encouragement*, according to its presented effects on employees, is assumed as an important determinant of employees' motivation in the context of Innovation Contests such as those that have already been proven in other or more general contexts. The corresponding hypothesis is as follows:

Hypothesis 2a: *A higher level of perceived supervisory encouragement is positively related to employees' motivation to participate in Innovation Contests.*

Regarding employees' *affective organizational commitment*, the existence of supervisor support as an important facet of the work environment with a positive influence is often discussed (cf. Eisenberger *et al.*, 2010; Rhoades *et al.*, 2001). This issue is summarized in the following quotation: "Employees who interpret the supervisor's caring and positive regard as coming from the organization should feel an obligation to return the caring and positive regard that results in an increased *affective organizational commitment*" (Eisenberger *et al.*, 2010, p. 1087). Eisenberger *et al.* (1990, p. 55) argue that "employees with high perceived support expressed greater affective attachment to the organization". Like the previous assumptions on *organizational encouragement*, a positive effect of *supervisory encouragement* on employees' *affective organizational commitment* is assumed in this study. Again, *affective organizational commitment* is seen as a general attitude with an organizational context of reference that is not specific to Innovation Contests. The formulated hypothesis reads as follows:

Hypothesis 2b: *A higher level of perceived supervisory encouragement is positively related to employees' affective organizational commitment.*

Regarding employees' *participation intention* and *supervisory encouragement,* the literature highlights the existence of "positive relationships between supportive supervision and employee in-role and extra-role performance" (Shanock and Eisenberger, 2006, p. 691). Furthermore, "research from the social and organizational support literature indicates that when supervisors are supportive of subordinates, this treatment leads to favorable outcomes for the employee and the organization such as reduced work stress and enhanced performance. [...] Such efforts would include enhanced performance of standard job activities, as well as helping behaviors that go beyond assigned responsibilities" (Shanock and Eisenberger, 2006, p. 690). Like the derivation of the hypotheses set forth above, in this study employees' performance is instantiated by extra-role performance, including employees' offers of voluntary suggestions, knowledge and skills (Chen *et al.,* 2009; Eisenberger *et al.,* 1990) and ideas for developing innovation within and for the organization (Shore and Wayne, 1993), all which are encouraged by Innovation Contests. On a related note, the dimension of internal communication, increased by the direct supervisor and fostered by Innovation Contests, has been shown to have a positive influence on innovative behavior (Foss *et al.,* 2011).

Taken together, as shown by similar studies in more general contexts, it is assumed that *supervisory encouragement* has a direct, positive effect on employees' extra-role performance, in that case expressed by their intention to participate in Innovation Contests. The corresponding hypothesis is as follows:

Hypothesis 2c: *A higher level of perceived supervisory encouragement is positively related to employees' intention to participate in Innovation Contests.*

3.2.2.3 The direct influences of organizational impediments

The negative influence of *organizational impediments* both generally and on (a) employees' *motivation*, (b) employees' *affective organizational commitment*, and (c) *participation intention in firm internal Innovation Contests* (Hypotheses H3a, H3b, H3c)

Organizational impediments are described as "an organizational culture that impedes creativity through internal political problems, harsh criticism of new ideas, destructive internal competition, an avoidance of risk, and an overemphasis on the status quo" (Amabile, 1997, p. 49). The consideration of *organizational impediments* focuses on a culture that impedes innovative behavior through internal politics, criticism, destructive competition, and risk avoidance (cf. Amabile, 1997, p. 49). In particular, the experiences of politics in the workplace have been discussed quite often (Gandz and Murray, 1980). In general, organizational politics and their perception are seen as "dysfunctional" for attitudes and work behavior (Cropanzano *et al.*, 1997). In that sense, LePine *et al.* (2005) have introduced the concept of hindrance stressors at work, "including demands such as organizational politics, red tape, role ambiguity" that consequently is "thwarting personal growth" (LePine *et al.*, 2005, p. 765).

Below, the relationships between *organizational impediments* and the corresponding dependent variables of *motivation, affective organizational commitment*, and *participation intention* are detailed.

Regarding employees' *motivation, organizational impediments* are described as follows: "Research suggests that internal strife, conservatism, and rigid, formal management structures within organizations will impede creativity. [...] Individuals are likely to perceive each of these factors as controlling, they may lead to [...] decreases in the intrinsic motivation that is necessary for creativity" (Amabile *et al.*, 1996, p. 1162). For instance, Kimberley and Evanisko (1981) have also highlighted that strong formal management structures lead to the impediment of creativity. Highlighting the stress-performance relationship, LePine *et al.* (2005, p. 764) have highlighted that one group of stressors at work is built from the so-called hindrance stressors, including, "e.g., role ambiguity, role conflict, hassles, red tape, etc.". They further state that "hindrance stressors should be associated with low motivation because people are not likely to believe that there is a relationship between effort expended coping with these demands and the likelihood of meeting them" (LePine *et al.*, 2005, p. 766). They also highlight the impact of conflicting role demands in the workplace, where employees "recognize that they cannot simultaneously satisfy both demands" and therefore "any effort expended to cope with the demands would likely be viewed as sapping resources that could otherwise be focused on demands associated with valued outcomes that could be met" (LePine *et al.*, 2005, p. 766). Bringing the context of this study to the fore, a conflicting role demand could occur in cases in which employees must choose between doing their day job or participating in Innovation Contests because of company-imposed conditions and circumstances. In summary, this study follows the previous findings regarding the effects of *organizational impediments* on employees' motivation, meaning that a higher level of impediments—i.e., politics, hindrance stressors, and role conflicts—results in a negative impact on employees' motivation to participate in firm internal Innovation Contests. The hypothesis reads as follows:

Hypothesis 3a: *A higher level of perceived organizational impediments is negatively related to employees' motivation to participate in Innovation Contests.*

Next, employees' *affective organizational commitment* and its influence as a result of *organizational impediments* are discussed. The literature describes perceived organizational politics in an organizational context as "unsanctioned influence attempts that seek to promote self-interest at the expense of organizational goals" (Randall *et al.*, 1999, p. 161). In their study, Randall and his colleagues analyze the influence of organizational politics on various forms of commitment, with the result that political perceptions are negatively related to *affective organizational commitment*. They argue that *affective organizational commitment* "is the extent to which the individual feels an emotional tie or bond to the organization. It was expected that individuals would form such ties with firms that are nonpolitical, because in the long run such organizations are most likely to meet their needs" (Randall *et al.*, 1999, p. 162). Similar findings are reported by Cropanzano *et al.* (1997). Additionally, Meyer *et al.* (2002) find that role ambiguity and role conflict are negatively correlated with *affective organizational commitment*. This study attempts to show a similar negative effect, thus, the next hypothesis reads as follows:

Hypothesis 3b: *A higher level of perceived organizational impediments is negatively related to employees' affective organizational commitment.*

Regarding employees' *participation intention* and *organizational impediments*, Randall *et al.* (1999, p. 161) argue that "political environments make for risky investments. For this reason, workers should attempt to contribute as little effort to the organization as is reasonably possible. Thus, politics should lead to lower performance". Considering the literature, organizational politics, leaders' behavior, and competition are highlighted as important: Cropanzano and his colleagues (1997) investigate the negative influence of perceived politics on desired work behavior, described as "positive work behaviors include such things as volunteering for extra work, […] actions at work which are above and beyond what is required of the employee. Based on our earlier observations, we expected that individuals are more likely to invest effort on behalf of the organization" (Cropanzano *et al.*, 1997, p. 162). According to Cropanzano *et al.* (1997, p. 162), "an individual who sees him or herself in a political setting has reason to belief that hard work will not be consistently rewarded. […] Additionally, the setting is potentially less predictable and more threatening. All of this should serve to make most individuals less happy and less apt to invest additional effort to maintaining the organization." Amabile *et al.* (2004) qualitatively investigates leaders' behavior in engaging creativity in the organization. They find several leader behaviors (cf. Amabile *et al.*, 2004, pp. 18ff.), including assigning work (e.g., creating time pressure, giving inappropriate assignments, failing to provide clarity), monitoring the status of work (e.g., checking work too often, displaying a lack of interest, providing nonconstructive feedback) and problem solving (avoiding problem solving, creating problems), are negatively related to perceived leaders' support and consequently to participation. The lack of clarity and lack of control over an individual's own ideas and an "unfair" idea evaluation are important negative influences on an employee's decision to participate in firm innovation, as assessed by Franke *et al.* (2013).

Moreover, the facet of competition is highlighted as follows: "If competition is perceived as threatening, as is often the case with in-group competition, creativity will tend to be affected negatively" (Amabile, 1988, pp. 149 f.) Focusing on communities for Innovation Contests, Hutter *et al.* (2011) highlight the role of a "climate for co-operation". The researchers introduce the term "community-based collaboration among competing contest participants", which stands for the co-existence of competitive and cooperative elements in communities and context settings. In line with Franke and Shah (2003), they argue that collaboration drastically decreases when competition is (too) high, e.g., because "community members become rivals, are competing against each other" (Hutter *et al.*, 2011, p. 5). Taken together, the perceptions of *organizational impediments* within the organization might attributable to the above-mentioned issue, negatively influencing employees' participation intention with respect to Innovation Contests. The hypothesis reads as follows:

Hypothesis 3c: *A higher level of perceived organizational impediments is neg-*
atively related to employees' intention to participate in Innova-
tion Contests.

3.2.2.4 The direct influences of workload pressure

The negative influence of workload pressure both generally and on (a) employees' *motivation*, (b) employees' *affective organizational commitment*, and (c) *participation intention* in firm internal Innovation Contests (Hypotheses H4a, H4b, H4c)

Workload pressure addresses a negative effect of excessive workload, which is perceived as a means of controlling individuals (cf. Amabile et al., 1996, p. 1161). Workload pressure is defined as "extreme time pressures, unrealistic expectations for productivity, and distractions from creative work" (Amabile, 1997, p. 49). With respect to the effects of workload pressure on creative tasks, it has been highlighted that "excessive workload pressure would be expected to undermine creativity, especially if that time pressure were perceived as imposed externally as means of control" (Amabile *et al.*, 1996, p. 1161). The aspect of time pressure is described as "insufficient time to think creatively about the problem; too great a workload within an unrealistic time frame; high frequency of 'fire-fighting'" (Amabile, 1988, p. 148). Furthermore, high workload can lead to psychological depletion (Walsh *et al.*, 2015).

Below, the relationships between *workload pressure* and the corresponding dependent variables of *motivation, affective organizational commitment*, and *participation intention* are detailed.

With respect to employees' *motivation* and *workload pressure,* causality is expressed by Walsh *et al.* (2015, p. 196) as follows: "High workload can lead to negative affective reactions (Gorgievski and Hobfoll, 2008), such as decreased motivation to perform job duties conscientiously". Similarly, the literature has described the influence of workload and time pressure on employees' emotional exhaustion (Houkes *et al.*, 2003; Lee and Ashforth, 1996). The researchers state, "demanding aspects of work (i.e., high workload) lead to constant overtaxing and in the long-term to exhaustion. [...] However, when stress ensues, and people experience stress every day, this may eventually result in a draining of one's energy and a state of emotional exhaustion" (Houkes *et al.*, 2003, p. 429). Formulated differently, "reducing levels of workload may prevent emotional exhaustion" (Houkes *et al.*, 2003, p. 446). Zhou and Shalley (2003), which asserts that close monitoring is negatively related to *motivation.* This study proposes, consistent with previous research, that *workload pressure* (including both time pressure and emotional exhaustion) has a negative influence on employees' *motivation* for participation in firm internal Innovation Contests. The corresponding hypothesis reads as follows:

Hypothesis 4a: *A higher level of perceived workload pressure is negatively related to employees' motivation to participate in Innovation Contests.*

Regarding employees' *affective organizational commitment* and the influence of *workload pressure*, again the connection is seen as independent of the Innovation Contest context (see explanation above). Walsh *et al.* (2015, p. 197) argue, "we forecast a negative impact of workload on organizational commitment [...]. Workload-induced resource depletion can activate employees' coping mechanism such that they purposely refuse to invest further effort [...]. Employees who decrease their investments tend to experience a corrosion of emotional attachment and reduced commitment toward the firm". In the literature on employees' emotional exhaustion, it is already well-known that exhaustion is related to both work overload and organizational commitment (cf. Lee and Ashforth, 1996). Additionally, Ahuja *et al.* (2002) report a negative influence of high workload on an individual's job satisfaction that ultimately results in a decrease in employees' organizational commitment. In summary, the following hypothesis is formulated to demonstrate the negative influence of *workload pressure* on employees' *affective organizational commitment* to the organization. Hypothesis 4b is as follows:

Hypothesis 4b: *A higher level of perceived workload pressure is negatively related to employees' affective organizational commitment.*

With respect to employees' *participation intention* and *workload pressure*, "work stress has an obvious negative impact on the individual and equally deleterious effects on the organization and the economy. The costs of stress can be enormous, due to lost time, reduced production" (Cropanzano *et al.*, 1997, p. 165). Therefore, people might concentrate on primary tasks instead of focusing on noncore tasks such as contributing to Innovation Contests: "High workloads can exhaust employees' mental and physical resources (Demerouti *et al.*, 2001) and induce negative organizational outcomes, such as neglecting noncore tasks [...] in an attempt to focus on core job tasks" (Walsh *et al.*, 2015, p. 196). Halbesleben and Bowler (2007) investigate the effect of emotional exhaustion on job performance and the mediating effects of motivation. They find that when "an employee is feeling exhausted because of his or her job (i.e., role overload), he or she would seek to protect resources" (Halbesleben and Bowler, 2007, p. 96). Regarding the consequences on extra-role performance, "it would not be expected that the employee would engage in behaviors that are not prescribed by his or her role, measured, or rewarded, as presumably that would require the very resources the employee is seeking to protect" (Halbesleben and Bowler, 2007, p. 96). Transferring these previous findings to firm internal Innovation Contests, this study assumes that *workload pressure* is related to employees' *participation intention* in a similar negative manner. The hypothesis reads as follows:

Hypothesis 4c: *A higher level of perceived workload pressure is negatively related to employees' intention to participate in Innovation Contests.*

3.2.2.5 The direct influences of employees' motivation

The positive influence of employees' *motivation* on their individual *intention to participate* in firm internal Innovation Contests (H5)

In general, "motivation is commonly considered a direct antecedent to performance" (Halbesleben and Bowler, 2007, p. 93). In similar articles, the literature has often highlighted the positive relationship between *motivation* and work-related behavior in an organizational setting (Pinder, 2014). Work *motivation* "has implications for the form, direction, intensity, and duration of behavior. This is, it explains what employees are motivated to accomplish, how they will attempt to accomplish it, how hard they will work to do so, and when they will stop" (Meyer *et al.*, 2004, p. 992). A further explanation of the relationships is as follows: "Task motivation makes the difference between what an engineer can do and what he will do. The former depends on his levels of expertise and creative thinking skills. However, it is task motivation that determines the extent to which he fully engage his expertise and creative thinking skills in the service of creative performance" (Amabile, 1997, p. 44). Additionally, "people who are primarily intrinsically motivated will be more likely to generate truly creative ideas" (Amabile, 1988, p. 143). Positive effects are not strictly limited to creativity; instead, "our findings were quite similar when we shifted our focus from creativity to the other elements of performance: Productivity, commitment to the work, and collegiality. [...] People performed better on all these fronts when there were in a good mood" (Amabile and Kramer, 2007, p. 9).

Additionally, studies in context similar to that of this thesis have reported a positive relationship between *motivation* and participation (Zheng *et al.*, 2011; Frey *et al.*, 2011; Leimeister *et al.*, 2009). Zheng *et al.* (2011) analyze the effects of task design and task attributes on participation in crowdsourcing contests. *Inter alia*, they find a positive connection between *motivation* and *participation intention*. Frey *et al.* (2011) show the relationship between participants' *motivation* and their contribution performance, arguing that intrinsic motivations are especially likely to increase the number of substantial postings and therefore, individuals with intrinsic enjoyment should be engaged for participation. Leimeister *et al.* (2009) highlight that the individual activation and *motivation* of participants are positively influenced by suitable functionalities of the underlying information technologies, leading to greater participation in IT-based idea competitions. Although the three aforementioned publications analyze Innovation Contests with external contributors, a similar positive effect between *motivation* and *participation intention* is hypothesized. To recap, hypothesis 5 displays the assumed positive relationship in the context of firm internal Innovation Contests:

Hypothesis 5:   *A higher level of employees' motivation is positively related to employees' intention to participate in Innovation Contests.*

3.2.2.6 The direct influences of employees' affective organizational commitment
The positive influence of employees' affective organizational commitment on their individual intention to participate in firm internal Innovation Contests (H6)

The positive influence of employees' *affective organizational commitment* on individuals' workplace behaviors has been shown in many organizational settings (cf. Mathieu and Zajac, 1990). Generally, a higher level of commitment and therefore a higher identification with the norms and goals of their employer leads to the greater likelihood of a display of behavior that favors the organization, which is described by Eisenberger *et al.* (2010, p. 1) as "an emotional attachment that fosters performance". Especially in the context of extra-role behaviors, Shore and Wayne (1993, p. 774) have shown either that *affective organizational commitment* is "important in explaining behaviors that are not formally rewarded or sanctioned by the organization, referred to as nonrole behaviors" or that "employees with high *affective organizational commitment* are motivated to help the organization reach its objectives via engagement in beneficial in-role and extra-role behaviors" (Eisenberger *et al.*, 2010, p. 5). One reason for this effect is described as follows: "Being committed to an organization generally means one feels proud of the organization and one's affiliation with it and is therefore glad to be a member of it. Those with high organizational commitment tend to be willing to put in extra effort for the organization" (Ahuja *et al.*, 2002, p. 2).

Although *affective commitment* to the organization is a focus of this study, a person's social affective identity with an online community is also considered important for engaging desired behaviors (Zhou, 2011): "Identification reflects an individual's identification with the community such as senses of belongingness and membership" (Zhou, 2011, pp. 69 f.). Zhou's results show that social identity and identification with the community has a significant influence on individuals' *participation intention*. Bateman *et al.* (2011) demonstrate that *affective commitment* positively influences the posting behavior of individuals who are part of a community. Focusing on crowdsourcing for firm innovation, Franke *et al.* (2013) highlight that users' ex-ante identification with the firm influences their individual fairness expectations, leading to an effect on their willingness and decision to participate in firm innovation activities. This work assumes similar effects on employees' decision to contribute to firm internal Innovation Contests. Taken together, although the relationship between employees' *affective organizational commitment* and their *intention to participate* in firm internal Innovation Contests has not yet been investigated, this study assumes a similar effect, as described above. In conclusion, hypothesis (H6) states as follows:

Hypothesis 6:   ***A higher level of employees' affective organizational commitment is positively related to employees' intention to participate in Innovation Contests.***

3.2.2.7   Indirect effects of positive work environment perceptions

In addition to the hypothesized direct effects, as presented in the previous sections, different moderation and mediation effects are predicted (cf. Baron and Kenny, 1986; Zhao *et al.*, 2010) and were considered in this study.

With respect to moderation effects, "a moderator is a qualitative (e.g., sex, race, class) or quantitative (e.g., level of reward) variable that affects the direction and/or strength of the relation between an independent or predictor variable and a dependent or criterion variable" (Baron and Kenny, 1986, p. 1174). In this study, positive work environment perceptions (*organizational encouragement* and *supervisory encouragement*) might function as moderators in the sense that the relations between *motivation* or *affective organizational commitment* and *participation intention* are influenced such that the effects are stronger for employees with higher positive perceptions of the work environment. *Organizational encouragement* and *supervisory encouragement* were chosen as moderator variables both because those variables were included in previous research (cf. Suh and Shin, 2008; Lin *et al.*, 2012; Duke *et al.*, 2009; Erdogan and Enders, 2007) and because these variables are of major interest in this thesis (see above). Some evidence of *organizational encouragement* as a moderating variable is available in the scientific literature. For instance, Suh and Shin (2008) investigate the moderation influence of *organizational encouragement* on the relation between "working hard" and "performance". In their study, the impact of *organizational encouragement* as a moderator is verified because "the findings provide evidence that the effect of working hard on performance can significantly be intensified when the practitioner is encouraged to work creatively in the organization" (Suh and Shin, 2008, p. 407).

In this study, similar effects of *organizational encouragement* as the moderator of the relations between both *motivation* and *participation intention* (model A) and *affective organizational commitment* and *participation intention* (model B), as pictured in Figure 14, are assumed. Regarding the magnitude of these effects, for employees the desired perception of *organizational encouragement* is assumed to strengthen the positive effects of *motivation* and *affective organizational commitment* on individuals' *participation intention* in firm internal Innovation Contests because employees perceive an increased organizational culture that encourages creativity through various aspects (cf. Amabile, 1997).

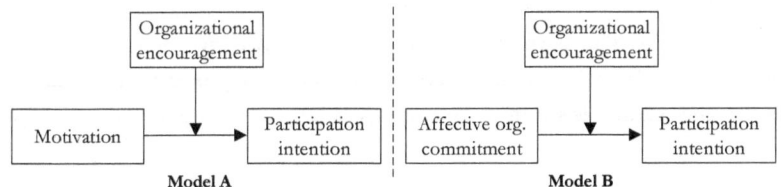

Figure 14: Models A and B for the moderation analysis[33]

The corresponding hypotheses (H7a and H7b) are set forth as follows:

Hypothesis 7a/7b: **The relations both among *motivation and participation intention* and among *affective organizational commitment* and *participation intention* are moderated by *organizational encouragement* such that the effect is stronger for employees who have a high perception of *organizational encouragement*.**

In a similar manner, the moderation effects of *supervisory encouragement* are hypothesized (see Figure 15) such that the relations between *motivation* and *participation intention* (model C) and between *affective organizational commitment* and *participation intention* (model D) are strengthened by the influence of *supervisory encouragement*. Here, *supervisory encouragement* might serve as part of an organizational culture that supports individuals, shows confidence in subordinate teams, and values individual contributions (Amabile, 1997).

---

[33] Author's own figure.

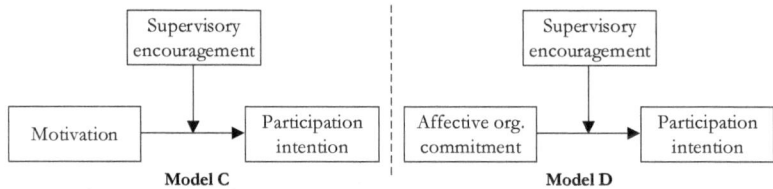

Figure 15: Models C and D for the moderation analysis[34]

Supervisory encouragement as a moderator has been analyzed by Lin *et al.* (2012). They have analyzed the moderating effect of *supervisory encouragement* (as part of the organizational culture) on organizational innovation. The researchers highlight that "when supervisors in the enterprise provide more encouragement, the degree of organizational innovation will be strengthened through the integration of an internal communication environment" (Lin *et al.*, 2012, p. 41).

The corresponding hypotheses (H8a and H8b) are set forth as follows:

Hypothesis 8a/8b: **The paths from *motivation and affective organizational commitment* to *participation intention* are moderated by *supervisory encouragement* such that the relation is stronger for employees high in their perception of *supervisory encouragement*.**

Regarding mediation effects, "a given variable may be said to function as a mediator to the extent that it accounts for the relation between the predictor and the criterion. [...] mediators speak to how or why such effects occur" (Baron and Kenny, 1986, p. 1176). In this study, mediation effects of *motivation* are assumed relevant in line with LePine *et al.* (2005, p. 766), who report an indirect effect of the work environment because it "should be indirectly [...] related to performance through motivation". Yeh-Yun Lin and Liu (2012) have analyzed the influence of *organizational encouragement* and *supervisory encouragement* on employees' performance by focusing on the mediating effect of work *motivation*; they state that "motivation in particular is seen as a crucial mediator of the relationship between climate and performance" (Yeh-Yun Lin and Liu, 2012, p. 56). A similar mediation effect is assumed in the context of firm internal innovation contexts (model A), as displayed in Figure 16. Moreover, *affective organizational commitment* is also hypothesized to mediate the relationship between *organizational encouragement* and *participation intention* (model B), similar to prior studies on organizational commitment as a mediator (cf. Yousef, 2000;

---

[34] Author's own figure.

Eisenberger *et al.*, 2010). Among others, Yousef (2000, p. 11) has stated that "organizational commitment mediates the relationships of leadership behavior with job satisfaction and job performance".

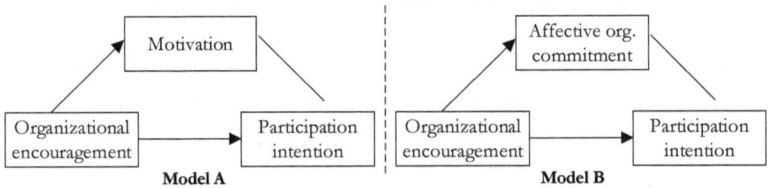

Figure 16: Models A and B for the mediation analysis[35]

The corresponding hypotheses (H9a and H9b) are set forth below:

Hypothesis 9a/9b: **The effect of *organizational encouragement* on *participation intention* will be mediated by *motivation/affective organizational commitment*.**

In a similar manner, the mediation effects of *supervisory encouragement* are hypothesized (see Figure 17) according to Yeh-Yun Lin and Liu (2012) for *motivation* (model C) and according to Raineri and Paillé (2016) for *affective organizational commitment* (model D). Yeh-Yun Lin and Liu (2012, p. 60) state that "employee's work motivation mediates the relationship between organizational creativity climate and perceived innovation". They verify a partial mediation effect of *supervisory encouragement*. Raineri and Paillé (2016, p. 7) highlight that "employee environmental commitment also mediates the positive relationship between supervisory support and environmental citizenship behavior".

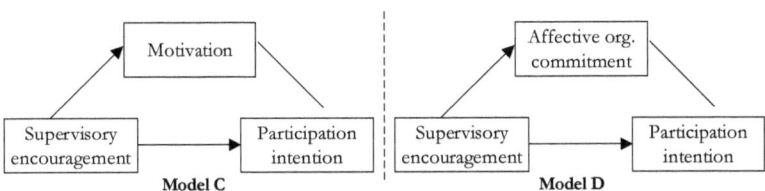

Figure 17: Models C and D for the mediation analyses[36]

---

[35] Author's own figure.
[36] Author's own figure.

The corresponding hypotheses (H10a and H10b) are described below:

Hypothesis 10a/10b: **The path from *supervisory encouragement* to *participation intention* will be mediated by *motivation/affective organizational commitment*.**

In addition to the detailed description of the relationships between the variables and the presentation of the formulated hypothesis, further details on data collection are provided below.

3.3 Data collection

In this section, the chosen measurements and the design of the web survey instrument are presented. Next, details of the sampling and data-collection procedures are presented.

3.3.1 Measurement model and design of the survey instrument

To develop the survey instrument, the (independent, moderating, and dependent) variables and corresponding scales are drawn from the existing literature. Table 14 represents all scales, number of items, rating options, and sources. All the constructs were measured using multiple, but at least 3, items (see Appendix C).

Scales	# of items	Ratings	Sources
Organizational encouragement	15	4-point work environment response scale	Amabile (2010), Amabile et al. (1996)
Supervisory encouragement	11	4-point work environment response scale	Amabile (2010), Amabile et al. (1996)
Organizational impediments	12	4-point work environment response scale	Amabile (2010), Amabile et al. (1996)
Workload pressure	5	4-point work environment response scale	Amabile (2010), Amabile et al. (1996)
Motivation	5	7-point-Likert scale	Zheng et al. (2011), Amabile et al. (1994)
Affective org. commitment	8	Affective commitment scale, 7-point-Likert scale	Allen and Meyer (1990)
Participation intention	3	7-point-Likert scale	Zheng et al. (2011)

Table 14: The measurement model[37]

---

[37] Author's own table.

For the work environment perceptions (as independent variables), the scales were adapted from Amabile (2010). All the scales are part of the "KEYS" instrument (former: The work environment inventory), which has been assessed as a reliable and valid instrument to investigate work environments for creativity (cf. Amabile et al., 1996), as claimed by the following quotation: "Data on KEYS gathered over a 12-year period, with over 12,000 individual employees from 26 different companies, have established the reliability and validity of this instrument" (Amabile, 1997, p. 47). For "*organizational encouragement*", the respondents describe the extent to which they agree with diverse statements related to encouraging an innovative work environment throughout their organization. For example, one representative item is as follows: "This organization has a good mechanism for encouraging and developing creative Ideas". Next, for "*supervisory encouragement*", respondents were asked to answer several questions about the encouragement of innovative behavior by their direct supervisor in their department, team, or project. For instance, one item was, "My supervisor values individual contributions to project(s)". During the adaption process, three items were excluded. Third, to assess "*organizational impediments*", respondents answered twelve items focusing on impediments of innovative behavior at work. One representative item reads as follows: "People in this organization are not very concerned about protecting their territory". Next, the fourth reason and aspect of work environment perceptions, "*workload pressure*", is assessed. Here, respondents should assess the level of how much their innovative work behavior is hindered by a high level of *workload pressure* both in their daily work and within their usual daily social and physical environment at work. One representative item reads as follows: "I do not have too much work to do in too little time". To measure all work environment perceptions, respondents used a four-point Likert-type scale, as suggested in the baseline article (Amabile *et al.*, 1995). A conscious decision to use 4-point measures was made because such a measure is the prerequisite for (i) a comparison with other results using the scales (Amabile *et al.*, 1996; Pirola-Merlo and Mann, 2004; Reid and De Brentani, 2012); and (ii) a comparison with quality criteria (i.e., the validity and reliability values) for the original scales (Amabile *et al.*, 1996). The scales ranged from 1= "never or almost never" to 4= "always or almost always". Work environment perceptions are assessed by the respondents in terms of their current work environment situation and therefore are assessed independently of the scenario of Innovation Contests.

    1 = Never or almost never true of your current work environment
    2 = Sometimes true of your current work environment
    3= Often true of your current work environment
    4 = Always or almost always true of your current work environment

For *motivation*, five items from Zheng et al. (2011) were adapted for use in the question-naire. After the introduction of the scenario, the participants rated their personal *motivation* for participation. One item reads: "I want to challenge myself to solve the problem in this Innovation Contest". For *affective organizational commitment*, all the items from the affective commitment scale (ACS) were chosen and adapted for this study (Allen and Meyer, 1990). Respondents rated the statements to indicate the extent to which their personal individual affective commitment to the organization is pronounced. For instance, one item was "I do feel "emotionally attached" to this organization". For *participation intention*, a scale re-ported by Zheng et al. (2011) was chosen. Respondents rated the level of their individual intention to participate in Innovation Contests if their company were to organize such activities in the future. Answers on the *motivation, affective organizational commitment* and the *participation intention* scale were given on a seven-point Likert scale anchored at 1= "strong-ly disagree" and 7= "strongly agree". In addition, several control variables were collected both for a detailed representation of the sample and for statistical reasons (see the follow-ing sections).

Table 15 gives an overview of the control and demographic variables.

Variables	Description
Gender	*The respondent's gender*
Age	*The respondent's age (in years)*
Tenure	*The respondent's tenure (in years)*
Education	*Selection field for highest educational status (e.g., vocational education, bachelor degree, master degree)*
Corporate function	*Selection field for the respondent's corporate function, (e.g., procurement, finance, HR)*
Occupation	*Selection field for the respondent's occupation (e.g., project manager, administrative staff)*

Table 15: List of control and demographical variables[38]

All the latent variables were designed as reflective measurement models (Coltman *et al.*, 2008), which is claimed to be a prerequisite for the following statistical analysis (cf. Chin, 1998; Roldán and Sánchez-Franco, 2012). Here, "covariation among measures is ex-plained by the variation in an underlying common latent factor" (Roldán and Sánchez-Franco, 2012, p. 197). In the composite, formative measurements models, "the measures

---

[38] Author's own table.

jointly influence the composite latent construct" (Roldán and Sánchez-Franco, 2012, p. 197). The scale level for all the scales of the measurement model is ordinal.

Figure 18 depicts the overall design of the questionnaire, which primarily follows the structure of the conceptual model.

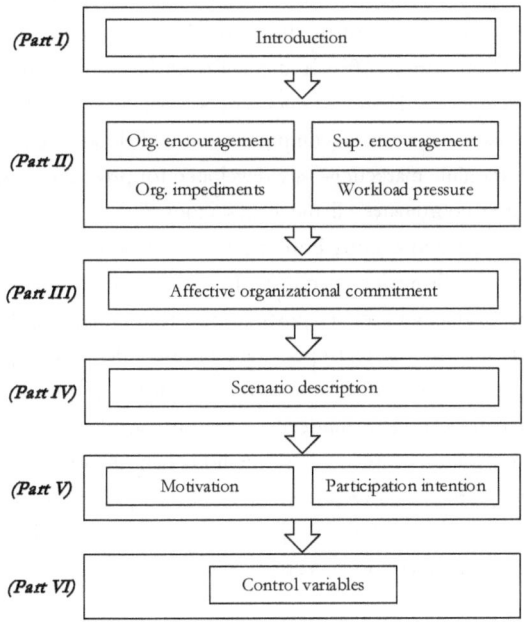

Figure 18: Structure of the questionnaire[39]

After a brief introduction (part I), responses regarding the four considered work environment perceptions are given in part II followed by the assessment of individuals' *affective commitment* to the organization (part III) and a short introduction of the scenario is presented (part IV). Next, responses for assessing respondents' *motivation* and *participation intention* are given (Part V). The questionnaire concludes with a request for diverse demographic and control variables in part VI.

---

[39] Author's own figure.

Next, further details for the selected sample and the data-gathering procedures are provided.

3.3.2 Sample and data-collection procedure

For this study, the survey was sent to employees of a large German DAX 30 company in the telecommunications industry all whom worked within a specific subunit responsible for the group-wide development of the firm's product and service portfolio. The unit is headed by the chief innovation officer and consists of the core area responsible for product and service innovations. There, products and services in several business segments—e.g., digital media, payment services, communication and cloud services, advertising business, television, and online marketing—are developed for private and business customers. In addition, the unit coordinates all the firm's innovation-oriented activities to exploit synergies different units and country borders. The unit coordinates the new product development efforts with firm-wide specialists. In addition, the subunit is responsible for the steering of the product and service portfolio and the entire group's product roadmap. To drive corporate growth, the unit opens new business fields and opportunities in line with corporate strategy. The subunit functions as the firm's "product house" and innovation laboratory and uses modern innovation and product development approaches, including Innovation Contests. Today, these methods help increase collaboration with internal and external partners and are accepted by all management levels. In its own statements, the subunit encourages the self-organization aspect of teams, leading to enhancements in both the development process and quality.

To reach the participants, several coordination activities and conversations with various firm managers were necessary prior to the data collection phase. In summary, more than 750 employees were reachable and were asked for their participation. For every completed questionnaire, one euro was donated to UNICEF. During the data-collection phase, steady monitoring of the participation and responses was performed. Before beginning the field period, the questionnaire was pretested using both a qualitative peer-review and a quantitative pretest with completed questionnaires for testing the scales and data-gathering mechanisms. More details of the data analyses procedures are presented below.

3.4 Data analysis

For the empirical analysis of the research model, the structural equation modeling (SEM) approach, supplemented by separate moderation and mediation analysis, was chosen. Structural equation modeling enables the "combined analysis of the measurement and the structural model" (Gefen *et al.*, 2000, p. 5) and testing of "the assumed causation among a set of dependent and independent constructs"; therefore, it is ideal for this study. The data analysis begins by analyzing the descriptive statistics of the sample (section 3.4.1), followed by an analysis of the scales (section 3.4.2).

3.4.1 Descriptive statistics

Data collection took place over ten weeks in autumn 2015 and was conducted in German. The call to participate in the online survey was sent to approximately 750 firm employees, resulting in 302 clicks on the link to start the questionnaire. Two hundred and thirty-one participants continued the survey after the welcome page; after data clearing (e.g., questionnaires that were excluded because they were incomplete or because the time to complete the survey was very low), 154 responses were used for data analysis (a response rate of approx. 20 %), a number that is a suitable sample size (Anderson and Gerbing, 1984). Iacobucci (2010, p. 92) notes that "a sample size of 100 will usually be sufficient for convergence, and a sample size of 150 will usually be sufficient for a convergent and proper solution".

The average duration of participation was 13.1 minutes. The age of the respondents was 39.1 years on average; women comprised 35.1 % of the group. The distribution of the respondents' age (see Table 16), educational status (see Table 17), job function (see Table 18) and occupation (see Table 19) are detailed in the corresponding tables. The descriptive statistics indicate a broad coverage of respondents from the surveyed unit. These findings are consonant with this study's aim in terms of reaching different people with different backgrounds during the firm-wide generation of novel ideas and innovations.

Age	Freq.	%
17-20	2	1.3 %
21-30	38	24.7 %
31-40	37	24.0 %
41-50	60	39.0 %
51-60	16	10.4 %
61-70	0	0.0 %
71-80	1	0.6 %
Total	**154**	**100.0 %**

Table 16: Respondents' age[40]

Education	Freq.	%
Secondary education	23	14.9 %
Vocational education	36	23.4 %
Matriculation standard	20	13.0 %
Bachelor degree	13	8.4 %
Master degree	54	35.1 %
PhD degree	6	3.9 %
Missing	2	1.3 %
Total	**154**	**100.0 %**

Table 17: Respondents' educational status[41]

---

[40] Author's own table.
[41] Author's own table.

Job function	Freq.	%
Sales, Marketing	24	15.6 %
IT	23	14.9 %
Finance	16	10.4 %
Production	14	9.1 %
R&D	10	6.5 %
Human Resources	7	4.5 %
Customer Care	6	3.9 %
Logistics	5	3.2 %
Procurement	2	1.3 %
Other	45	29.2 %
Missing	2	1.3 %
Total	**154**	**100.0 %**

Table 18: Respondents' job function[42]

Occupation	Freq.	%
Management	12	7.8 %
Head of Department	11	7.1 %
Team leader	21	13.6 %
Project manager	17	11.0 %
Technical staff	40	26.0 %
Administrative staff	49	31.8 %
Missing	4	2.6 %
Total	**154**	**100.0 %**

Table 19: Respondents' occupation[43]

As part of quality management, two questions were integrated into the survey. The first addresses the overall quality of the survey: "How do you evaluate the survey overall?" The second addresses the items: "Are all the questions clearly and understandably described?". Both of these questions were anchored on a five-point Likert scale ranging from 1 to 5. For the former question, the results show a "rather good" evaluation (mean: 2.15, standard derivation: 0.748, cumulated percent of "1= very good" and "2= rather good" rankings: 73 %). For the latter question, the results show even better results (mean: 1.86, standard derivation: 0.705, cumulated percent of "1= fully agree" and "2= agree": 85 %).

3.4.2 Analysis of scales and constructs

Following the observation of the descriptive statistics, the analysis of the scales and constructs is performed. Therefore, both an exploratory factor analysis and a confirmatory factor analysis are executed with different goals, as explained in the following sections.

3.4.2.1 Exploratory factor analysis

On the one hand, the fixed assignment of the indicators to the latent variables is given from the theories that are used in this study; nonetheless, an exploratory factor analysis was executed to show the dimensionality and number of factors that are included in the measured constructs (Weiber and Muehlhaus, 2014). In pursuit of this goal, the identification and exclusion of indicators that might not be a consequence of the latent variable, as required by a reflective measurement model (cf. Coltman *et al.*, 2008), are performed. As a

---

[42] Author's own table.
[43] Author's own table.

second requirement, the existence of a high correlation between the items (Coltman *et al.*, 2008) is validated.

In the first step, for the group of items with one latent variable, the Kaiser-Meyer-Olkin-Criteria (KMO) and the Bartlett Test specify the togetherness of the variables (see Table 20 for the results of the EFA). In this study, the KMO values for all the scales are greater than the required cutoff value of 0.6 (cf. Kaiser and Rice, 1974) and the Bartlett test with significance levels of 0.000 shows a good suitability for the factor analysis (cf. Dziuban and Shirkey, 1974). For the factor extraction, the principal components analysis (PCA) is used because PCA "is the default method of extraction" (Costello and Osborne, 2005, p. 1) and is used in more than the half of the studies that Costello and Osborne (2005) consider in their meta-study. According to the Kaiser criteria, extracted factors with eigenvalues larger than 1 are considered (Weiber and Muehlhaus, 2014). For rotation, following Weiber and Muehlhaus (2014), "promax" rotation is chosen. A content-related interpretation of the extracted factors is done at the end of the factor analysis with the results that were available at the time.

Scales	KMO values	Explained variance	Factor 1 eigenvalue	Factor 2 eigenvalue	Factor correlation
Organizational encouragement	0.925	53.36 %	6.868	1.137	0.664
Supervisory encouragement	0.921	68.76 %	5.501	-	-
Organizational impediments	0.891	53.72 %	5.150	1.297	0.592
Workload pressure	0.715	71.55 %	2.432	1.146	0.306
Affective org. commitment	0.871	68.74 %	4.368	1.131	0.009
Motivation	0.793	61.57 %	3.079	-	-
Participation intention	0.766	90.36 %	2.711	-	-

Table 20: Results of the exploratory factor analysis[44]

In the second step, the measure of sampling adequacy (MSA) and the commonalities of the items are checked (cf. Weiber and Muehlhaus, 2014). The MSA value specifies how strong an item must be seen as belonging with the other items. With respect to common-

---

[44] Author's own table.

alities, the values specify the percentage of the variables distribution that is explained by the extracted factor. For both checks, indicators with values of less than 0.5 are excluded from further research steps. Below, the peculiarities regarding the scales and the EFA are presented. A complete representation of the results, including a verification of every indicator for the EFA, is included in Appendix A.

For *organizational encouragement*, the analysis reveals a solution with two factors that have eigenvalues of 6.868 and 1.137 and an explained variance of the two factors (after rotation) of 53.36 %. The correlation between the two factors is 0.664. Four indicators ("rewarding creativity", "mechanisms", "risk taking", "handling unusual ideas") were excluded from further analysis because their commonality (0.439/0.493/0.197/0.308) was below the cut-off of 0.5 (cf. Weiber and Muehlhaus, 2014, p. 132). For *supervisory encouragement*, a one-factor solution with an eigenvalue of 5.501 and an explained variance of 68.764 % was identified. For *organizational impediments*, the analysis reveals a solution with two factors that have eigenvalues of 5.150 and 1.297 and an explained variance of the two factors of 53.72 %. The correlation between the two factors is 0.592. The indicators of "TMT[45] risk taking" and "change emphasis" are excluded because their commonalities (0.250/0.351) are below the cut-off value of 0.5. For *workload pressure*, the analysis reveals a solution with two factors that have an eigenvalue of 2.432 and 1.146 and an explained variance of 71.55 %. The correlation between the factors is 0.306. For *affective organizational commitment*, the analysis reveals a solution with two factors that have an eigenvalue of 4.368 and 1.131 and an explained variance of 68.74 %. The correlation between the factors is 0.009. The item "discussing with outside people" (commonality value is 0.470) and the item "attachment to other firms" (MSA value is 0.450) are excluded. For *motivation*, the analysis reveals a solution with one factor that has an eigenvalue of 3.079 and an explained variance of 61.57 %. Again, two items ("like the things I do" and "making new experiences") are excluded because their commonality values (0.490/0.488) did not reach the cut-off value of 0.5. Finally, for *participation intention*, the analysis reveals a solution with one factor that has an eigenvalue of 2.711 and an explained variance of 90.362 %.

Below, the reliability of the factors and indicators were tested. With respect to the suitability of the indicators for measuring the constructs, some issues occur: For *organizational impediments (factor 1)*, the corrected item-scale correlation shows that the item "strict control" has a value of 0.485, which is below the cut-off value of 0.5 and therefore excluded. In a similar vein, both items for *workload pressure (factor 2)* are excluded because both of the

---

[45] TMT= Top Management Team

values for the corrected item-scale correlation are 0.421 and the Cronbach's alpha value is only 0.582 (therefore, the scale is completely excluded from the next steps). Finally, with respect to *affective organizational commitment* and in a similar vein, the item "organization's problems" is excluded because Cronbach's alpha again increases (0.902 to 0.911). For all scales with changes and exclusions, the calculation of the reliability of the indicators was repeated until the results indicated reliability (Churchill, 1979). Thus, in the *motivation* scale, the item "curiosity" (increase in Cronbach's alpha from 0.848 to 0.878 was also excluded. Table 21 shows the Cronbach's alpha values (all greater than 0.778) and the inter-item correlations (specifying the average correlation between the indicators) that are all greater than the cut-off value of 0.3 (cf. Weiber and Muehlhaus, 2014, p. 138), whereas for the sake of clarity, the corrected item-scale correlations (for each indicator) are not mentioned.

Scales	Cronbach's alpha	Inter-item-correla-
Organizational encouragement (factor 1)	*0.871*	*0.493*
Organizational encouragement (factor 2)	*0.803*	*0.507*
Supervisory encouragement	*0.935*	*0.642*
Organizational impediments (factor 1)	*0.778*	*0.471*
Organizational impediments (factor 2)	*0.859*	*0.554*
Workload pressure (factor 1)	*0.778*	*0.537*
Affective organizational commitment	*0.911*	*0.670*
Motivation	*0.878*	*0.651*
Participation intention	*0.947*	*0.855*
*Workload pressure (factor 2)**	*0.582*	*0.421*

Note: [1]Mean values, *Scale excluded from further analysis because of low corrected item-scale correlations between indicators

Table 21: Construct reliabilities[46]

3.4.2.2 Confirmatory Factor Analysis

Subsequently, validation of the measurement model is performed to proof the reliability and validity of the latent variables. Therefore, a confirmatory factor analysis was conducted to examine the quality of the measurement model. The CFA reveals some issues with the indicator reliability, which is below the cut-off value of 0.5 (see Appendix B for details), as described below. With respect to *organizational encouragement (factor 1)*, two items ("idea evaluation" and "failure acceptance") are excluded (see Appendix B for values). In addition, another two items ("problem solving" and "expectations") of *organizational en-*

---

[46] Author's own table.

couragement (factor 2) are excluded. Regarding *organizational impediments*, two items ("work pressure", "formal procedures") are excluded. Pursuant to Kline's (2011) suggestion that in CFA, each factor should be measured by at least two indicators, "idea criticism" remains in the scale (although indicator reliability is 0.477, which is less than 0.5). For *workload pressure*, one item ("distractions") is excluded. Similar to the EFA, the complete representation of the results of the CFA is presented in Appendix B. Following the CFA, the following structure for the latent variables is revealed (see Table 22). Finally, the discriminant validity is tested as the final step before analyzing the hypothesized causalities.

Scales	Factors	Factor reliability	# of indicators	Indicators
Organizational encouragement	*Factor 1*	*0.872*	*5*	*Enthusiasm, Performance evaluation, Atmosphere, Idea flow, Shared vision*
	Factor 2	*0.797*	*2*	*Support of new ideas, Recognition*
Supervisory encouragement	*Factor 1*	*0.935*	*8*	*Expectations, Planning, Goal setting, Communication, Interpersonal skills, Confidence, Valuing contributions, Serves as good example*
Organizational impediments	*Factor 1*	*0.766*	*2*	*Work criticism, Idea criticism*
	Factor 2	*0.861*	*5*	*Political problems, Destructive competition, Protecting territories, Project hindering, Destructive criticism*
Workload pressure	*Factor 1*	*0.782*	*2*	*Workload, Time pressure*
Affective org. commitment	*Factor 1*	*0.913*	*5*	*Spend rest of career, Part of the family, Emotionally attached, Personal meaning, Sense of belonging*
Motivation	*Factor 1*	*0.879*	*2*	*Challenging oneself, Figuring out own proficiency*
Participation intention	*Factor 1*	*0.947*	*3*	*Intention to participate, Attempt to participate, Decision to participate*

Note: Number of indicators reported without excluded items because of low reliability (less than 0.5)

Table 22: Results after the preliminary confirmatory factor analysis[47]

3.4.2.3 Test of discriminant validity and common method variance

Next, to analyze for discriminant validity (Zait and Bertea, 2011; Henseler *et al.*, 2015; Farrell and Rudd, 2009), Table 23 provides an overview of studied variables and correlations between the latent factors. All the correlations show consistent values both with respect to their direction and in comparison to former research. The calculated average variance extracted (AVE) for each scale is more than 0.5, as required in the literature (Chin, 1998; Fornell and Larcker, 1981). Discriminant analysis (by comparing the square root of the

---

[47] Author's own table.

AVE with the absolute values of the correlations with other variables) revealed two validity concerns: The first issue involves *organizational encouragement (factor 1)* and *organizational encouragement (factor 2)*. The second issue involves *organizational impediments (factor 1)* and *organizational impediments (factor 2)*.

Latent variables	AVE	1	2	3	4	5	6	7	8	9
Organizational encouragement 1	0.529	**0.727**[a]								
Organizational encouragement 2	0.589	0.963[b*]	**0.768**							
Supervisory encouragement	0.644	0.549	0.522	**0.802**						
Organizational impediments 1	0.502	-0.596	-0.462	-0.133	**0.709**					
Organizational impediments 2	0.554	-0.409	-0.235	-0.177	0.826[*]	**0.744**				
Workload pressure	0.619	-0.038	-0.187	-0.040	0.426	0.447	**0.787**			
Motivation	0.784	0.229	0.356	0.104	-0.005	0.006	-0.011	**0.886**		
Affective org. commitment	0.679	0.495	0.406	0.411	-0.174	-0.128	0.062	0.142	**0.824**	
Participation intention	0.857	0.220	0.378	0.061	-0.036	0.008	0.115	0.851	0.263	**0.926**

Note: [a]Diagonal elements are square root AVE; [b]Inter-construct correlations
[*]Validity concerns (square root of AVE is less than correlation)

Table 23: Results of the discriminant validity check[48]

In light of the results of the discriminant validity check, both of the factors for *organizational encouragement* and *organizational impediments* are joined into a single latent variable each. Next, a renewed CFA was conducted (see Table 25). The items "performance evaluation" and "atmosphere" (from *organizational encouragement*) and the items "political problems", "idea criticism", and "work criticism" (from *organizational impediments*) were excluded because of low indicator reliability. The renewed values for the factor reliabilities are 0.890 for *organizational encouragement* and 0.874 for *organizational impediments*. Finally, the renewed discriminant validity check shows no validity concerns (see Table 24).

---

[48] Author's own table.

Latent variables	AVE	1	2	3	4	5	6	7
Organizational encouragement	0.555	**0.744**[a]						
Supervisory encouragement	0.644	0.542[b]	**0.802**					
Organizational impediments	0.553	-0.341	-0.183	**0.743**				
Workload pressure	0.625	-0.109	-0.041	0.444	**0.790**			
Motivation	0.784	0.304	0.104	-0.017	-0.007	**0.886**		
Affective org. commitment	0.679	0.449	0.411	-0.147	0.055	0.144	**0.824**	
Participation intention	0.857	0.304	0.063	-0.005	0.113	0.852	0.265	**0.926**

Note: [a]Diagonal elements are square root AVE; [b]Inter-construct correlations

Table 24: Results of the final discriminant validity check[49]

Comparing the means of the work environment perceptions, the means in this study are between the mean values reported for high- and low-creativity projects except for *organizational encouragement* (cf. Amabile *et al.*, 1996, p. 1175); *organizational encouragement* (high-creativity: 2.83/low-creativity: 2.51); *supervisory encouragement* (high: 3.12/low: 2.78); *organizational impediments* (high: 2.05/low: 2.32); and *workload pressure* (high: 2.52/low: 2.62).

---

[49] Author's own table.

Scales	Indicators	Factor load-ings	Factor reliability	AVE
Organizational encouragement	Support of new Ideas	0.772		
	Enthusiasm	0.749		
	Recognition	0.715	0.862	0.555
	Idea flow	0.769		
	Shared vision	0.717		
Supervisory encouragement	Expectations	0.784		
	Planning	0.755		
	Goal setting	0.728		
	Communication	0.839	0.935	0.644
	Interpersonal skills	0.804		
	Confidence	0.804		
	Valuing contributions	0.811		
	Serves as good example	0.882		
Organizational impediments	Destructive competition	0.762		
	Protecting territories	0.757	0.832	0.553
	Project hindering	0.708		
	Destructive criticism	0.746		
Workload pressure	Workload	0.718	0.768	0.625
	Time pressure	0.857		
Affective organizational commitment	Spend rest of career	0.708		
	Part of the family	0.793		
	Emotionally attached	0.901	0.913	0.679
	Personal meaning	0.807		
	Sense of belonging	0.896		
Motivation	Challenging oneself	0.891	0.879	0.785
	Figuring out own profi-	0.880		
Participation intention	Intention to participate	0.937		
	Attempt to participate	0.942	0.947	0.857
	Decision to participate	0.898		

Table 25: Results after the final confirmatory factor analysis[50]

---

[50] Author's own table.

The model fit indices of the final confirmatory factor analysis show good values, as indicated by $\chi^2/df=$ 1.519, comparative fit index (CFI)= 0.924, root mean square error of approximation (RMSEA)= 0.058, and the standardized root mean square residual (SRMR)= 0.0615.

Because common method bias could be a problem in this study given that all the variables are collected from the same respondents, appropriate tests are performed. Following Podsakoff *et al.* (2003), various approaches were used during the post-collection statistical analysis to detect common method variance (CMV). Accordingly, the Harman's single factor test (SPSS-Options: Extraction – *Fixed Number of factors= 1*, Rotation – Method: None) shows that only one single factor explains a minority of the explained variance (29 %) that is clearly below the critical value of 50 %. Additionally, the test that adds a common latent factor (CLF, using IBM SPSS AMOS 23) to a model with all other latent variables and their indicators included in the analysis. By comparing the standardized regression weights of the two models both with and without the CLF (that is correlated with all observed variables in the model), no large differences could be discovered. The largest difference is 0.152, which is not greater than 0.2, therefore, the data are not biased. In summary, common method bias is not a problem in this study.

3.5 Presentation of empirical findings

3.5.1 Validation of the research model

In this section, a validation of the research model is presented. Therefore, a co-variance-based structural equation modeling was conducted in line with former research (e.g., Wang and Tsai, 2014; Suh and Shin, 2005) to test the hypotheses model. The path diagram was designed using IBM SPSS AMOS 23 with maximum likelihood (ML) as the estimation procedure due to an only moderate violation of the normal distribution[51]. AMOS uses covariance-based SEM (CBSEM), compared to variance-based SEM, similar to, for instance, the partial least squares (PLS) approach supported by SmartPLS (Roldán and Sánchez-Franco, 2012). The choice between variance and covariance-based SEM should be done in a conscious way and in this study, "in situations where prior theory is strong and further testing and development are the goal, covariance-based full-

---

[51] The assessment of normality in the AMOS Output shows that only for the kurtosis of "Protecting territories" (-1.045) and for "Serves as good example" (-1.015) a value larger than 1 exists. Additionally, only the C.R. values for the skew-indices of "Confidence", "Sense of belonging", and "Intention to participate" are larger than the threshold of 2.57. In total, a moderate violation of the normal distribution can be assumed according to Weiber and Muehlhaus (2014, pp. 180 f.).

information estimation methods are more appropriate" (Roldán and Sánchez-Franco, 2012, p. 201).

To report the model fit indices, the recommendations of Kline (2011) are followed, including the suggestion to report the Chi-square/df value, the comparative fit index (CFI), the root mean square error of approximation (RMSEA), and the standardized root mean square residual (SRMR). To interpret the model fit indices, Hooper *et al.*'s (2008) guidelines for assessing indices were considered (see Table 26).

Indices	$\chi 2$/d.f.	CFI	RMSEA	SRMR
Value	*1.382*	*0.954*	*0.050*	*0.088*
Cut-Off	*< 2.000*	*> 0.900*	*< 0.080*	*< 0.500*

Table 26: The model fit indices[52]

Here, $\chi 2$/df must be low as 2.0, the CFI value should have a value greater than 0.9, RMSEA's upper limit must be less than 0.08 and the SRMR should be less than 0.5 (cf. Hooper *et al.*, 2008). The assessment of the model fit indicates a good fit between the model and the data, as shown by the considered model fit indices (following the application of modification indices, see below in this section).

Focusing on interpreting the results of the structural model, Table 27 summarizes the variables, hypotheses and paths of that model. Furthermore, the standardized path estimates, their significance levels, standard errors (S.E.) and critical ratios (C.R.) are provided. For the standard errors, the values should be at a similar level and as low as possible to indicate a reliable parameter estimation (cf. Weiber and Muehlhaus, 2014, p. 229), whereas the values in this study are between 0.078 and 0.226. All the C.R. values show a value of 1.96 or above, indicating that the estimated parameters differ significantly from zero (Weiber and Muehlhaus, 2014). To interpret the influences among the latent variables, seven of the fourteen hypotheses could be confirmed after assessing the direction and significance of the effects.

---

[52] Author's own table.

Dep. var.	Hypo.	Paths	Estimates[1]	S.E.	C.R.
Affective or-ganizational commitment	**H1a**	*Org. encouragement → Affect. org. com.*	**0.298****	**0.177**	**2.88**
	H2a	*Sup. encouragement → Affect. org. com.*	**0.234***	**0.123**	**2.42**
	H3a	*Org. impediments → Affect. org. com.*	-0.046$^{n.s.}$	-	-
	H4a	*Workload pressure → Affect. org. com.*	**0.171***	**0.168**	**2.22**
Motivation	**H1b**	*Org. encouragement → Motivation*	**0.378*****	**0.226**	**3.33**
	H2b	*Sup. encouragement → Motivation*	-0.089$^{n.s.}$	-	-
	H3b	*Org. impediments → Motivation*	0.096$^{n.s.}$	-	-
	H4b	*Workload pressure → Motivation*	-0.053$^{n.s.}$	-	-
Participation intention	H1c	*Org. encouragement → Part. intention*	0.049$^{n.s.}$	-	-
	H2c	*Sup. encouragement → Part. intention*	-0.107$^{n.s.}$	-	-
	H3c	*Org. impediments → Part. intention*	-0.003$^{n.s.}$	-	-
	H4c	*Workload pressure → Part. intention*	**0.102***	**0.144**	**1.96**
	H5	*Motivation → Part. intention*	**0.834*****	**0.078**	**11.73**
	H6	*Affect. org. com.→ Part. intention*	**0.154***	**0.079**	**2.47**

Significance level: * p< 0.05 / ** p< 0.01 / *** p< 0.001
Note: n.s.= not significant / [1]standardized

Table 27: Path estimates of structural model[53]

First, the results show a positive relationship between *organizational encouragement* and *affective organizational commitment* (H1a, β= 0.298, p< 0.01); *supervisory encouragement* is also positively related to *affective organizational commitment* (H2a, β= 0.234, p< 0.05). Therefore, hypotheses H1a and H2a are supported. Additionally, *workload pressure* shows a positive relationship to *affective organizational commitment* (H4a, β= 0.171, p< 0.05), whereas H4a could not be supported predicting a negative influence. Moreover, hypothesis H1b could be proved, meaning that *organizational encouragement* is also positively related to motivation (H1b, β= 0.378, p< 0.001). As the only work environment perception, *workload pressure* shows a positive direct influence on *participation intention* (H5, β= 0.834, p< 0.001). Finally, *motivation* shows a positive direct effect on *participation intention* (H5, β= 0.834, p< 0.001) and *affective organizational commitment* is positively related to *participation intention* (H6, β= 0.154, p< 0.05). The other relationships do not show significant results. Overall, three direct effects on *affective organizational commitment*, one direct effect on *motivation*, and three direct effects on *participation intention* are revealed (see Figure 19).

---

[53] Author's own table.

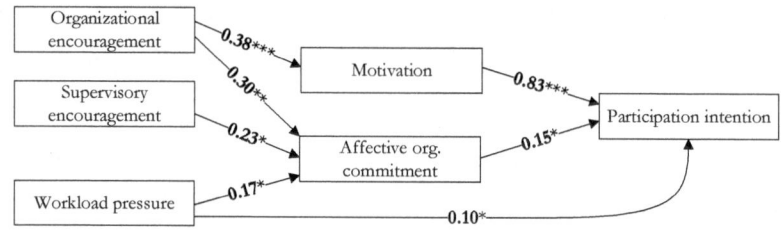

Figure 19: Significant standardized effects between latent variables[54]

Regarding the squared multiple correlations (SMC), which indicate the percentage of the variance of the three dependent variables that can be explained by the other latent variables, the value for *participation intention* is 76.3 %, the value for *affective organizational commitment* is 24.9 % and the value for *motivation* is 12.7 %.

Based on considering the modification indices for a systematic improvement of the goodness of fit, two covariances between items' error terms were drawn after a content-related verification (instead of, for instance, removing the items from the model). The baseline was a model fit with acceptable indices: χ^2/df= 1.663, CFI= 0.920, RMSEA= 0.066, SRMR= 0.141. First, in the *supervisory encouragement* scale, the error terms of the items "expectations" and "goal setting" were correlated because of the very high modification index (38.49). This connection is confirmed, as highlighted by Tesluk *et al.* (1997, p.33): "A strong goal emphasis can also help facilitate the develop of self-efficacy expectations regarding their own creative behavior". The model fit indices improve to χ^2/df= 1.550, CFI= 0.934, RMSEA= 0.060, SRMR= 0.141) In addition, in the *affective organizational commitment* scale, the error term of the item "part of the family" is correlated with "sense of belonging" because of a high modification index (17.77). Again, this connection shows a togetherness because both items are part of the affective commitment scale (Allen and Meyer, 1990). The improved model fit indicates χ^2/df= 1.487, CFI= 0.942, RMSEA= 0.056, and SRMR= 0.142. Additionally, according to the recommendations of Weiber and Muehlhaus (2014, p. 249), a covariance between the latent variables *organizational encouragement* and *supervisory encouragement* was drawn (high modification index: 34.08). The final model fit indices (as mentioned above) are χ^2/df= 1.382, CFI= 0.954, RMSEA= 0.050, and SRMR= 0.088.

Considering control variables, several methods and recommendations for the integration of control variables into organizational research have been used in the past (Becker,

---

[54] Author's own figure.

2005). In this study, the observed control variables of *age, tenure,* and *gender* are included[55] in a separate model, as suggested by Becker (2005, p.286): "Run and report the primary results both with and without the MCVs[56]". These results show significant relationships between *age* and *participation intention* (β= -0.164, p< 0.01) and between *tenure* and *participation intention* (β= 0.157, p< 0.01). In contrast, *gender* did not show a significant influence on *participation intention.* The model fit indices only show marginal changes (χ^2/df= 1.371, CFI= 0.947, RMSEA= 0.049, SRMR= 0.0895). Regarding the structural model's estimates and the primary results, all the relationships were essentially identical: Only the relationship between *affective organizational commitment* and *participation intention* becomes insignificant because of the integration of the control variables.

Subsequently, the additionally executed moderation and mediation analysis and the corresponding results are presented in line with the above-formulated hypothesis.

3.5.2 Moderation analysis

As hypothesized a priori at the conceptualization stage of this study, several moderation analyses are additionally executed based on the significant positive relationships between *motivation* and *participation intention* and between *affective organizational commitment* and *participation intention.* In pursuit of this goal, a moderation analysis (Baron and Kenny, 1986) using linear regressions and the extension module PROCESS 2.13.2 for SPSS (Hayes, 2013) is executed. Prior to the calculation, the independent and moderator variables were mean centered.

First, with respect to the moderation analysis for the relationship from *motivation* to *participation intention,* with *organizational encouragement* as the moderator (H7a), the resulting regression analysis revealed an explained variance of 62 % in *participation intention.* Whereas *motivation* (independent variable) shows significance (b= 0.79, p< 0.01), *organizational encouragement* as moderator and the interaction term does not show significance. Therefore, the absence of a moderation effect of *organizational encouragement* is indicated.

Second, with respect to the relationship between *affective organizational commitment* (independent variable) and *participation intention* (dependent variable), a model that included *organizational encouragement* as moderator variable was tested (H7b). The moderation analysis was calculated using PROCESS and explained approximately 12 % of the variance in *participation intention.* The results reveal that *affective organizational commitment* (b= 0.23, p< 0.05) and *organizational encouragement* (b= 0.41, p< 0.05) have a significant influence on *participa-*

---

[55] "Age" and "tenure" are covaried.
[56] MCVs= Measured Control Variables.

tion intention. Additionally, the interaction term of *affective organizational commitment* and *organizational encouragement* has a significant effect (b= 0.23, p< 0.05) on the outcome variable *participation intention* that confirms the existence of a moderation effect.

Analyzing the moderation effects in more detail, the recommendations of Spiller *et al.* (2013) were followed to investigate various regions of the moderator. For medium and high levels (+1SD) of *organizational encouragement*, the relationships between *affective organizational commitment* and *participation intention* are significant ((b_{medium}= 0.23, p< 0.05 and b_{high}= 0.38, p< 0.01). It is only for a low level (-1SD) of *organizational encouragement* that the effect of *affective organizational commitment* on *participation intention* does not show significance. Next, the simple slopes (see Figure 20) were drawn to visualize the effects (Aiken and West, 1991). With respect to the effects of *affective organizational commitment* at a high level (plotted at the right half of the figure), the results show significantly higher values for *participation intention* if *organizational encouragement* is high compared to a low level of *organizational encouragement*. When compared in this manner, considering *affective organizational commitment* at a low level (plotted at the left half), the effect of *organizational encouragement* (high or low) as a moderator variable shows no difference for the effect on *participation intention*.

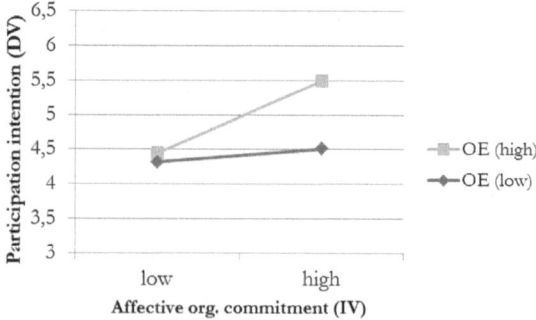

Figure 20: Slope analysis and explanation of moderation effect (for Model B)[57]

In addition, the moderation analyses with *motivation* (H8a) and *affective organizational commitment* (H8b) as independent variables, *supervisory encouragement* as the moderator and *participation intention* as the dependent variable do not show significant results.

---

[57] Author's own figure.

Summarizing the moderation analysis, it can be revealed that there is a significant direct positive effect from *motivation* to *participation intention* (see structural model for influence in general), and the relation is not moderated by *organizational encouragement* or *supervisory encouragement*. Furthermore, the effect from *affective organizational commitment* to *participation intention* is moderated by *organizational encouragement*, but not by *supervisory encouragement*.

3.5.3 Mediation analysis

3.5.3.1 Motivation as mediator variable

As the final part of the empirical analysis, various mediation analyses were conducted based on the detailed relationships and influences, as presented in section 3.2.2. First, a model with *organizational encouragement* as the independent variable, *motivation* as the mediator variable and *participation intention* as the dependent variable was analyzed (H9a), again with the PROCESS macro for SPSS.

The results (see Table 28) show that *organizational encouragement* has a direct effect on *motivation* (b= 0.51, p< 0.01) but not a direct effect on *participation intention* (b= 0.16, n.s.) and that these results are similar to the findings in the structural model. Additionally, *motivation* shows a direct significant influence on *participation intention* (b= 0.81, p< 0.001). In conclusion and according to Zhao *et al.* (2010), the support of an "indirect-only" mediation effect is shown. The direct, indirect and total effects are set forth below. For the indirect effect, the bootstrap confidence interval (between "lower level for confidence interval" (LLCI) and "upper level for confidence interval" (ULCI)) does not comprise zero (LLCI= 0.148; ULCI= 0.682), thus demonstrating the significance of the indirect effect (Preacher and Hayes, 2004).

Paths	Estimates	Sign.	R^2
Organizational encouragement → Motivation	*0.51*	**	*7%*
Organizational encouragement → Participation intention	*0.16*	*n.s.*	*62%*
Motivation → Participation intention	*0.81*	***	
Direct effect	*0.16*	*n.s.*	
Indirect effect	*0.41*	*yes*	
Total effect	*0.57*	***	
Significance level: * p< 0.05 / ** p< 0.01 / *** p< 0.001 Note: n.s.=not significant			

Table 28: Results of mediation analysis (*motivation* as mediator)[58]

In addition, a model with *supervisory encouragement* as the independent variable, *motivation* as the mediator and *participation intention* as the dependent variable (H10a) do not support a mediation effect. The results indicate that *supervisory encouragement* has no significant effect on *motivation* (b= 0.16, n.s.) and no significant effect on *participation intention* (b= -0.29, n.s.). Only the effect from *motivation* to *participation intention* was significant (b= 0.83, p< 0.001).

3.5.3.2 *Affective organizational commitment as mediator variable*

Next, a model with *organizational encouragement* as the independent variable, *affective organizational commitment* as the mediator and *participation intention* as the dependent variable was analyzed with PROCESS (H9b). The results indicate that *organizational encouragement* influences *affective organizational commitment* (b= 0.82, p< 0.001) and affects *participation intention* (b= 0.44, p< 0.05). Finally, the effect from *affective organizational commitment* to *participation intention* was not significant, indicating the absence of a mediation effect. In that case, the direct effects previously identified in the structural model seem to be essential.

Finally, a model with *supervisory encouragement* as the independent variable, *affective organizational commitment* as the mediator and *participation intention* as the dependent variable (H10b) was analyzed using PROCESS (see Table 29). The results show that *supervisory encouragement* has a direct effect on *affective organizational commitment* (b= 0.67, p< 0.01) but not on *participation intention* (b= -0.07, n.s.) and that these results are similar to the findings in the structural model. Additionally, *affective organizational commitment* shows a direct significant influence on *participation intention* (b= 0.27, p< 0.005), resulting in support of an "indirect-only" mediation effect (Zhao *et al.*, 2010). The direct, indirect and total effects are also set forth below. For the indirect effect, the bootstrap confidence interval does not include

---

[58] Author's own table.

zero (LLCI= 0.060; ULCI= 0.371), thus attesting to the significance of the indirect effect (Preacher and Hayes, 2004).

Paths	Estimates	Sign.	R²
Supervisory encouragement → Affective organizational com-	*0.67*	*****	*14 %*
Supervisory encouragement → Participation intention	*-0.07*	*n.s.*	*6 %*
Affective organizational commitment→ Participation inten-	*0.27*	****	
Direct effect	*-0.07*	*n.s.*	
Indirect effect	*0.18*	*yes*	
Total effect	*0.11*	*n.s.*	

Significance level: * p< 0.05 / ** p< 0.01 / ***p< 0.001
Note: n.s.= not significant

Table 29: Results of mediation analysis (*affective organizational commitment* as mediator)[59]

Regarding the achieved results for the mediation effects, it can be revealed that the effect from *organizational encouragement* to *participation intention* is mediated by *motivation* (as formulated in Hypothesis 9a) and the effect from *supervisory encouragement* to *participation intention* is mediated by *affective organizational commitment* (as formulated in hypothesis 10b).

Finally, the summary and conclusion for this study are presented in the final section of this chapter.

---

[59] Author's own table.

3.6 Summary of Study A

This first study *("STUDY A: The work environment and participation in Innovation Contests")* and its outcomes provide evidence that employees' perceptions of their work environment have a significant influence on their *affective organizational commitment,* their *motivation* and their *participation intention* in firm internal Innovation Contests. Through the use of a web survey instrument, with data collected in cooperation with a German DAX 30 company, the statistical analysis reveals several causalities. Specifically, and among other identified relationships, *organizational encouragement* impacts both employees' *motivation* and their *affective organizational commitment.* Additionally, *supervisory encouragement* influences individuals' *affective organizational commitment.* Furthermore, it is validated that both employees' *motivation* and their *affective organizational commitment* significantly influences their *participation intention.* Moreover, the existence of further indirect effects of positive work environment perceptions can be revealed.

A detailed discussion of the results, along with their theoretical and practical implications and avenues for further research, are provided in chapter 5.

4 STUDY B:

Exploring an engaging work environment for Innovation Contests

In the last chapter, several work environment perceptions as important determinants of Innovation Contest success could be verified. In particular, the existence of encouragement and support both by the entire organization (*organizational encouragement*) and by the direct superior (*supervisory encouragement*) were shown to have a strong influence on both employees' *motivation* and their *affective organizational commitment* as important personal attitudes and finally, on employees' *participation intention* for firm internal Innovation Contests. Based on these findings, the aim of Study B is not only to delve deep into the phenomenon of an engaging work environment that creates organizational support for firm internal Innovation Contests but also to extend existing knowledge because to date, the areas of action for the organizational support of firm internal Innovation Contests have not been systematically investigated. This study complements the findings of the first quantitative study, which verifies important causalities. Study B mitigates one disadvantage of quantitative research, i.e., that only the facts that are directly addressed by the survey questions are covered by the research. Therefore, a qualitative investigation involving the conceptualization and execution of a qualitative content analysis is executed to identify, structure and present the best practices condensed from the experiences of 18 large companies in Germany, Great Britain, France, Finland, the Netherlands, Switzerland, and Norway.

This chapter is structured as follows: First, the definition of the study's aim and overall research design is provided in section 4.1. Next, the selected scenario and the sampling strategy are introduced in section 4.2. Subsequently, the data collection (in section 4.3) and data analysis techniques (in section 4.4) are detailed, followed by the presentation of the results in section 1.1. Finally, a summary of the study and the findings is presented in section 0.

4.1 Aim and study design

In this section, the aim of the study is presented in section 4.1.1. Next, the study design and the adherence to quality criteria are illustrated in section 4.1.2.

4.1.1 Aim of the study

As illustrated in the introduction, no consolidated view of the constituents of organizational support for firm internal Innovation Contests exists in the scientific literature. Therefore, the aim of Study B (as formulated in the first chapter)

"… is an identification and in-depth analysis of the areas of action of an engaging work environment in the context of firm internal Innovation Contests (exploratory study design). Therefore, a cross-case comparison of best practices and a comparison with the relevant body of knowledge are executed (Section 1.5, "Research design").

Research question 2 (see section 1.2 on "Derivation of the research questions") reads as follows: "Which areas of action (critical success factors, constituents and facets) exist that represent aspects of an engaging work environment in firm internal Innovation Contests?"

As shown in the theoretical background (chapter 2), these aspects of an engaging work environment as social-environmental influences might commonly shape employees' work environment perceptions and ultimately, their work-related behavior in terms of creativity and innovation (Amabile *et al.*, 1996). The existence and strength of positive perceptions is crucial and "the level at which the source of influence operates is less important than the perceptions themselves […]. For example, whether individuals feel their co-workers, their supervisors, or their high-level superiors encourage them to take risk in their project work, what is important is the fact that they perceive such encouragement" (Amabile *et al.*, 1996, p. 1157).

In this study, a central question that has existed in the research for many years is now addressed in the context of Innovation Contests: "How can environmental factors in organizations promote motivation and creativity?" (Amabile, 1988, p. 146). Accordingly, this study has "with the goal of exploring, explaining and predicting a phenomenon in a real-world context, namely, the existence of an engaging work environment that supports the success of Innovation Contests" (Section 1.5, "Research design"). A qualitative content analysis, as my chosen research method, enables "rendering the rich meaning associated with organizational documents combined with powerful quantitative analysis" (Duriau *et al.*, 2007, p. 7). Through a structured data collection and analysis process, it is possible to go "deep into the phenomenon". In the subsequent step, a "simplification of the phenomenon" can be reached by gradually increasing the abstraction level. The characteristics of this exploratory study are described as follows (see Table 30).

Properties	Criteria
Nature of study	*Exploratory, inductive reason-*
Study design	*Qualitative*
Research method(s)	*Content analysis*

Table 30: Overview of the study design for Study B[60]

This study's goal is to reveal the areas of action in an engaging work environment for firm internal Innovation Contests on different levels: Critical success factors on the highest level, constituents on the subjacent level, and facets and representative actions on the concrete level (see later in this chapter for detailed introduction). In comparison, several studies investigate various factors that are critical to success when examining the innovation management literature (e.g., Cooper and Kleinschmidt, 1995, 2007). Cooper and his colleague investigate the question, "what are the critical success factors that underlie excellent new product performance?" (Cooper and Kleinschmidt, 2007, p. 1). They identify the factors for determining the profitability, impact, and performance of new product development efforts. Their "study was designed to uncover the drivers of performance", and they "were able to obtain much greater insights into the factors and practices that really discriminate between the top and poor performers" (Cooper and Kleinschmidt, 2007, pp. 52–53). Cooper and Kleinschmidt's work had a very high impact in the community, as shown by the fact that it is one of the six most frequently referenced articles in the Research Technology Management Journal (Cooper and Kleinschmidt, 2007) and the fact that it is published in the well-known Journal of Product Innovation Management (Cooper and Kleinschmidt, 1995). Moreover, the identification of critical success factors for different issues has been completed in a broad range of different domains, for example, ERP implementations (Somers and Nelson, 2001), software development projects (Ahimbisibwe *et al.*, 2015), new product development (Cooper and Kleinschmidt, 2007, 1995), and crowd-sourcing projects (Luettgens *et al.*, 2014).

Investigating and identifying a consolidated and entire view of areas of action in an engaging work environment for firm internal Innovation Contests might be a complex issue because in the actual state of the art, there has been no consolidated representation of the

---

[60] Author's own table.

various dimensions and relevant aspects are described in articles spread across a broad range of publications and research fields (cf. Adamczyk *et al.*, 2012). This exploratory study investigates (1) the critical success factors, subjacent constituents, and facets of an engaging work environment and comprehensive organizational support based on practical experiences on both the provider side and the customer side; and (2) the development and consolidated representation of an integrated view. The data analysis reveals detailed insights provided directly by innovation experts working every day on the utilization of firm internal Innovation Contests embedded in an innovation initiative. In conclusion, this study's formulated aim might be strengthened by the following quote: "There are also some success factors of Innovation Contests organizers should follow when conducting such a contest" (Adamczyk *et al.*, 2012, p. 344). Details on the study design and adherence to the quality criteria are presented in the next section.

4.1.2 Study design and quality criteria

As mentioned above, a qualitative content analysis following Mayring (2014) is conducted both to identify all aspects of the issue and to answer the research question. A suitable design for a qualitative content analysis involves comparing similar phenomena inferred from various bodies of text (cf. Krippendorff, 2004, p. 93). Here, the same type of content analysis must be applied to each part within a body of texts. To do so, "there is a need for researchers to have a strategy that guides them in the execution of the research" (Maimbo and Pervan, 2005, p. 1281). Following this important requirement, all important aspects of the study design, the adherence to quality criteria, and the sampling strategy were defined a priori for the study execution. The conceptualization and preliminary findings of this study were presented at the annual Open and User Innovation Society Meeting (Hoeber, 2015). The study design is presented in detail in section 4.1.2.1. Subsequently, the quality criteria and their adherence are introduced in section 4.1.2.2.

4.1.2.1 Study design

As a study design, the step-by-step model for qualitative research (see Table 31) is adapted from Mayring (2014, pp. 53 f.). The first ("concrete research question") and second steps ("linking research question to theory") have already been established within the framing of this thesis. The "definition of the research design" (step 3) is presented in this section. It is followed by the "definition of the sampling strategy" (step 4) and "methods of data collection and analysis" (step 5). The "presentation of results" is described (step 6) thoroughly with regard to the existing literature. Finally, the "discussion of quality criteria" (step 7) is provided in the summary. Detailed information about how the various research steps are designed is given in the referenced chapters and sections.

Steps	Description	Consideration in ...
Step 1	*Concrete research question*	*Chapter 1*
Step 2	*Linking research question to theory*	*Chapters 1* and *2*
Step 3	*Definition of the research design*	*Section 4.1.2 (this section)*
Step 4	*Definition of the sampling strategy*	*Section 4.2*
Step 5	*Methods of data collection and analysis*	*Sections 4.3 (data collection)* and *4.4 (data analysis)*
Step 6	*Presentation of results*	*Section 4.5*
Step 7	*Discussion of quality criteria*	*Sections 4.1.2 (this section)* and *4.6*

Table 31: Steps of the research process for Study B[61]

Furthermore, details about the applicability and the design (either explorative or descriptive) of qualitative content analysis might be considered in a completely defined study design, as presented below.

The applicability of qualitative content analysis is described as follows: "Content analysis, a class of methods at the intersection of the qualitative and quantitative traditions, is promising for rigorous exploration of many important but difficult-to-study issues of interest to management researchers" (Duriau *et al.*, 2007, p. 5). The authors determine a broad range of fields of application across the subdomains of management research, including strategy, managerial cognition, organizational behavior, human resources (HR), technology management and others that have been used during the last 25 years (cf. Duriau *et al.*, 2007, p. 8). In general, content analysis is assessed as a suitable technique for operationalizing expert knowledge (Krippendorff, 2004, p. 90).

One important difference of qualitative studies is the design, which is either explorative or descriptive (Mayring, 2014). The descriptive design, which is characterized by a deductive approach, is characterized by "working through the texts with a deductively formulated category system" (Mayring, 2014, p. 12) and the aim "to filter out particular aspects of the material [...] according to pre-determined ordering criteria or to assess the material according to certain criteria" (Mayring, 2014, p. 64). Discussing the study design in this case, one opportunity for a deductive approach would be to adopt the aspects that influence *organizational encouragement* as described by Amabile *et al.* (1996, pp. 1159 f.). These dimensions are first, "encouragement of risk taking and of idea generation"; second, "fair, sup-

---

[61] Author's own table, referencing Mayring (2014, p. 15).

portive evaluation of new ideas"; third, "reward and recognition of creativity"; and fourth, "collaborative idea flow across an organization and participative management and decision making". Calling back to Amabile's differentiation and comparing it with the aims of this study, Amabile's differentiation is assessed as too broad and abstract because a more granular investigation is required here. Consequently, the existing dimensions as described by Teresa M. Amabile are not used in this study because those researchers focus on work environment perceptions in general, whereas this study is tailored to Innovation Contests.

In contrast, the explorative design, which is characterized by an inductive category development, is described as "formulating new categories out of the material" and as a suitable approach that "reduced the material in such a way that the essential contents remain, in order to create through abstraction a comprehensive overview of the base material which is nevertheless still an image of it" (Mayring, 2014, p. 64). One argument for such open approaches, compared to predefined categories, is to "demand not to block the open sight on the subject by theories" (Mayring, 2014, p. 11). Mayring (2014, p. 79) summarizes the specialties of an inductive category formation. With respect to the filtering of relevant information, "only those parts relevant for specific research questions are considered. For this selection process a rule of selection is formulated". With respect to category development, the aim "is to arrive at summarizing categories directly, which are coming from the material itself, not from theoretical considerations." He further highlights that the "category definition is a central step in content analysis, a very sensitive process. […] It aims at a true description without bias owing to the preconceptions of the researcher, an understanding of the material in terms of the material."

This study aims to develop aspects and categories that are as close as possible to the data (Mayring, 2004). In conclusion, an explorative design combined with an inductive approach is chosen with respect to the research question because to the best of my knowledge, no suitable category system is available in the pertinent literature. Therefore, the chosen study design is similar to the approach of grounded theory (Strauss and Corbin, 1990; Strauss, 1987), in which the process of forming categories from the data is described as "open coding" (see Mayring, 2014, p. 79).

4.1.2.2 Quality criteria

Adherence to quality criteria is an important quality to prove when engaging in content analysis (Miles *et al.*, 2013; Miles and Huberman, 1984) and has been considered from the outset of this study as described as follows. In line with Miles *et al.*'s (2013, pp. 311 ff.) discussion of drawing valid meanings from content analysis, along with their formulation of "standards for the quality of conclusions", the quality criteria as illustrated in Table 32 are considered in this study.

Quality Criteria	Description[93]
Objectivity/ Confirmability	"The basic issue can be framed as one of relative neutrality and reasonable freedom from unacknowledged researcher biases—at the minimum, explicitness about the inevitable biases that exist"
Reliability/ Dependability	"The process of the study is consistent, reasonably stable over time and across researchers and methods. We are addressing issues of quality and integrity: Have things been done with reasonable care?"
Internal validity/ Credibility	"The crunch question: Truth value. Do the findings of the study make sense? Are they credible to the people we study and to our readers? Do we have an authentic portrait of what we were looking at?"
External validity/ Transferability	"We need to know whether the conclusions of the study [...] have any larger import. Are they transferable to other contexts? Do they fit? How far can they be generalized?"
Utilization/ Application	"Even if a study's findings are valid and transferable, we still need to know what the study does for its participants–both researchers and researched–and for its customers."

Table 32: Quality criteria for Study B[62]

Below, details about how the study design complies with the various criteria are presented. **Objectivity/confirmability** is described "as total independence of the research results from the researcher, is held to be difficult within qualitative approaches" (Mayring, 2014, p. 14). Nevertheless, this criterion could be strengthened by the following suggestions, which are considered in this study. Miles *et al.* (2013, pp. 311 f.) highlight that to ensure that "methods and procedures are described explicitly and in detail", readers "can follow the actual sequence of how data were collected, processed, condensed/transformed and displayed" and "the researcher has been explicit and as self-aware as possible about personal assumptions". To increase the transparency of this criterion, all the details regarding the data collection and analysis are described in this thesis. In addition, the corresponding records, data set, and codings are documented in specialized software MAXQDA, which was used throughout the analysis process. "MAXQDA [...]

---

[62] Author's own table, referencing Miles *et al.* (2013, pp. 311 ff.).

can be used for coding, accessing text, displaying the completed codes, writing memos and presenting the results in the form of tables and graphs" (Oliveira *et al.*, 2013, p. 305). Therefore, all data and information are "available for reanalysis by others" (Miles *et al.*, 2013, p. 312).

To ensure **reliability/dependability**, Mayring (2014, p. 10) proposes to "formulate strict content-analytical rules for the whole process and for the specific steps of analysis. [...] The Qualitative content analysis itself is to be understood as a data analysis technique within a rule guided research process". Additionally, Mayring (2014, p. 14) highlights that to increase reliability/dependability "within qualitative content analysis, the rule guided procedures can strengthen this criterion". Consistent with this requirement, the aim and the research design, the different research steps and the analysis rules and procedures are defined in a precise manner (see section 4.4.1 for a detailed description). In line with Shenton (2004), reliability/dependability is increased by focusing on company experts for firm internal Innovation Contests as key informants for studying the topic and obtaining a deep understanding of the phenomenon.

Regarding **internal validity/credibility**, Mayring (2014, p. 14) states that "validity in a broader sense is usually less of a problem within qualitative approaches, because they seek to be subject centered, close to everyday life". To ensure internal validity/credibility, the study design considers data triangulation, e.g., through the integration of various sources of evidence. Additionally, showing direct quotations from the data strengthens the meaning of the findings. Miles *et al.* (2013, p. 313) describe this quality criterion, *inter alia*, in a manner such that the result "makes sense, seems convincing or plausible" and the "findings are clear, coherent". Additionally, this work received acceptance and positive feedback from the expert community running the OUI Workshop with respect to a research-in-progress publication of this study (see Hoeber, 2015).

Regarding **external validity/transferability**, it is important to clarify how easily a transfer of the outcome can be made to other contexts. This quality could be strengthened both by providing a detailed description of this study's underlying scenario and by presenting the sampling strategy and selected cases. To increase the transferability of this study, the design considers comparing the results with the body of literature and previous findings, especially during the presentation of the identified critical success factors and the results.

Finally, the **utilization/application** of the findings is increased through the formulation of practical and theoretical implications that help us understand how this study and the achieved results can be used in future activities, both in theory and in practice. It is especially important both that "the level of usable knowledge offered is worthwhile, ranging from consciousness raising and the development of insight or self-understanding to broader considerations" and that "the findings are intellectually and physically accessible to potential users" (Miles *et al.*, 2013, p. 315).

Subsequently, the chosen scenario and details for the sampling strategy are presented.

4.2 Scenario and sampling

After introducing the aim and study design, the selected scenario forming the context for this study is presented in the section 4.2.1. The sampling strategy is illustrated in section 4.2.2.

4.2.1 Scenario

As the scenario for this study, I chose one of the leading software development and consulting companies specializing in Innovation Contest software that supports customer-companies' innovation-management processes. The provider-company was founded in 2001, inspired by dissatisfaction with the status quo in the innovation practices in the firm that then employed the founders. Over the last fifteen years, the provider has successfully implemented and supported more than 200 active customers, with installations for corporate Innovation Contests worldwide and across all industries (CEO Provider Company[63]). Today, the utilization of the software solution is spread over several branches, e.g., Logistics, Automotive, Chemicals, Engineering and Aviation. For instance, customers include the following multinational companies: Siemens, BASF, P&G, Deutsche Post DHL, Bosch, Bombardier, General Electric and Deutsche Bahn. The software provider's portfolio offers full-lifecycle innovation management to its customers to cover the entire innovation process in a single solution (see Figure 21).

---

[63] Presentation at the annual Innovation Managers Forum 2015.

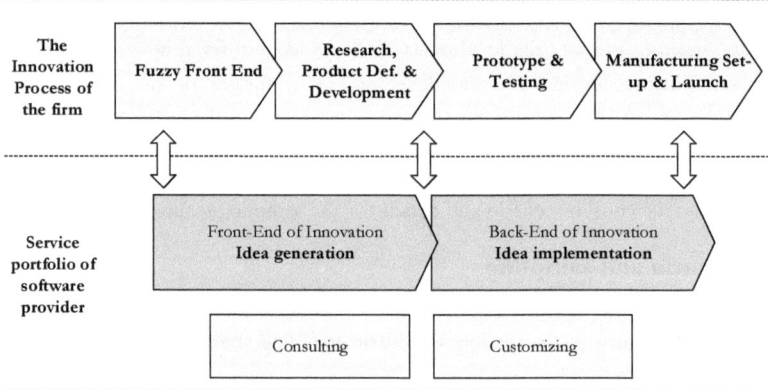

Figure 21: The software provider's solution portfolio[64]

The software systemizes all phases of the innovation process and supports collecting, improving, evaluating and selecting innovative ideas. The process is separated into several target-oriented phases, all them based on Innovation Contests in defined focus areas. At the front end of innovation, "strategic innovation areas" (SIA) are defined and derived based on the corporate strategy and establish the target areas for ideas. Next, "Innovation Contests" are efficient means for generating ideas among a heterogeneous group of employees. Ideas are developed and discussed in the context of the Innovation Contests and based on relevant quality criteria. At the back end of innovation, ideas can be developed into "concepts" and "implementation projects". Concepts are created based on selected ideas by extending a business case and preparing the project realization. The implementation of the best ideas is based on well-founded decisions.

The software supports these important steps with time-saving management functions for classifying, evaluating and selecting the most promising ideas. As service offerings, consulting and customizing services are offered. A specific department and permanent contacts provide a wide range of consultancy services that support clients in all steps of the implementation. The carefully elaborated customizing possibilities of the platform enable the design of individualized forms, processes, events and actions, and access rights primarily through an easy-handing configuration interface. The provider is convinced that employees' motivation is a decisive factor in the program's success. The core product is a web-based application for the end-to-end management of the innovation process with comprehensive coverage of innovation-management activities (see Figure 22).

---

[64] Author's own figure, innovation process taken from McNally *et al.* (2011, p. 67).

The solution is conceptualized for an implementation and customization into a specific organizational environment; it can be customized for individual requirements and circumstances. The innovation-management platform can work as a standalone solution or can integrate into corporations' existing social software solutions, e.g., IBM Connections or YIVE. To further investigate and represent an integrated perspective on organizational support as the phenomenon of interest, this technological innovation serves as the initial point of entry.

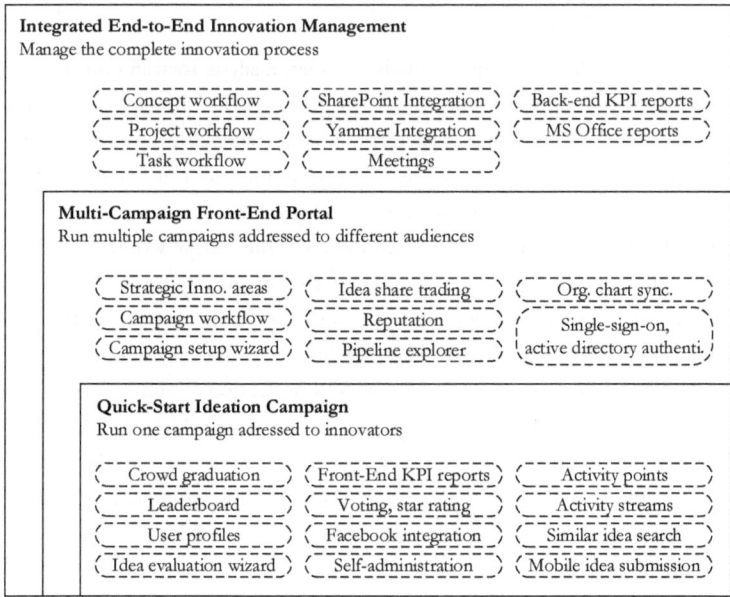

Figure 22: Product features of the Innovation Contests solution[65]

Next, the sampling strategy for defining the group of cases chosen in this study is introduced.

4.2.2 Sampling strategy

Because the study design has an exploratory nature (see section 4.1.2), diverse cases were selected that are likely to be representative in the sense of covering the variation among the relevant population (Seawright and Gerring, 2008). Duriau *et al.* (2007) call for "multiple sources of information [...] when new and unique phenomena [...] are studied. [...]

---

[65] Author's own figure, following the vendor's product documentation.

This approach ensures the validity of the research through data triangulation and the incorporation of the perspectives from multiple participants" (Duriau *et al.*, 2007, pp. 17–18). To minimize selection bias, a relatively large group of cases is chosen (see below).

To address the research question through an adequate sampling strategy, the following criteria must be fulfilled by the practice partners: (1) a focus on intra-organizational (and e.g., not inter-organizational) Innovation Contests; (2) an Innovation Contest approach and a corresponding web-based platform that are already in place and in a mature state within the customer companies; and (3) a willingness to participate in this study.

For this study, a rich basis for the qualitative content analysis focusing on the phenomenon of organizational support in the context of this thesis is formed by the best-practice presentations that different branches gave at Germany's annual Innovation Managers Forum. The forum is one of the biggest conferences on innovation management in Germany and offers an exchange of insights, trends, and practical experiences presented by client-firms that have already implemented firm internal Innovation Contests. For two days each, presentations by customers from different regions (mainly across Europe) and the provider firm were presented, adopting a retrospective perspective. The official description by the provider[66] is as follows: "The [...] Innovation Managers Forum is the place where [...] clients come together with innovation professionals and leading industry experts from around the world to share, develop, and get inspired by new ideas. Our event is dedicated to serving innovation professionals and tackling the challenges they face". Furthermore, "the [...] Innovation Managers Forum [...] is an intense, rewarding 2-day conference, which focuses on business-critical and industry-relevant topics related to innovation and idea management. The [...] client community and industry experts share insights about their experiences, discuss challenges and propose solutions, and evaluate new innovation trends. Customers come away from the forum fueled with new ideas, inspired by the experience of fellow peers, and prepared to enhance their innovation programs within their organizations."

All the contributions from the annual Innovation Managers Forums in 2013, 2014, and 2015 were considered. This group forms the foundation both for this analysis and for investigating the phenomenon. Further details related to the data-collection procedures are presented below.

---

[66] This information is taken from publicly available marketing materials.

4.3 Data collection

4.3.1 Sources of evidence

As noted in the conceptualization section of this study, data triangulation is achieved by integrating several sources of evidence in the data collection and analysis process. For that reason, best-practice presentations from 18 different European firms (Germany, Great Britain, France, Finland, the Netherlands, Switzerland, and Norway) were included. Table 33 gives an overview of the cases, including industry and country of residence.

The number of employees, which range from 3,700 to more than 400,000, is consistent with the desired context of this study, that is, large, multi-divisional firms. The last column illustrates the role of the innovation experts (e.g., Director Group Innovation, Head of Idea and Innovation Management, Innovation Program Manager) that were included in the study. The experts were considered key informants (Marshall, 1996; Kumar et al., 1993): Because of their function and job characteristics, "a key informant is an expert source of information" and furthermore, "key informants, as a result of their personal skills, […] are able to provide more information and a deeper insight into what is going on around them" (Marshall, 1996, p. 92).

No	Case	Industry	Category	Country
1	A	ICT	Provider	GER
2	B	Construction#	Customer	FIN
3	C	ICT	Customer	GER
4	D	Logistics#	Customer	GER
5	E	Automotive#	Customer	FRA
6	F	Chemicals#	Customer	GER
7	G	Insurance*,#	Customer	GER
8	H	Automotive*,#	Customer	GER
9	I	Aviation	Customer	GER
10	J	Environmental Services	Customer	FRA
11	K	Chemicals	Customer	GER
12	L	ICT	Customer	NET
13	M	Chemicals	Customer	SWI
14	N	Engineering	Customer	GER
15	O	Production	Customer	GER
16	P	Consulting	Customer	SWI
17	Q	Oil & Gas	Customer	NOR
18	R	ICT*	Customer	GBR

* = Use of Innovation Contests within subsidiary,

\# = Only slides are available (no videos)

Table 33: Units of investigation (Provider and customers)

In some cases, more presentations or different informants could be considered because different contributions were made to the Innovation Managers Forum in the studied years. Only a minimum number of best-practice presentations were excluded from this study because they did not fit into the sample of large, multi-divisional firms, e.g. because they represent research institutes, universities, or innovation intermediaries. In addition to the collection and preparation of best-practice presentations, several company documents, Webinars and white papers published by the software provider were considered in this study but not explicitly integrated in the coding procedures. This information both reflects the experiences of the provider company and offers a rich source of interesting additional insights.

Further details on the data gathering and preparation procedures are provided in the next section.

4.3.2 Data gathering and preparation

In general, various types of material can be analyzed using the content analysis technique. Following Mayring (2014), a content analysis instrument can analyze a broad range of texts (e.g., newspapers, media, protocols, documentation, charts, web pages). Additionally, Krippendorff states that the "content analysis is a research technique for making replicable and valid inferences from texts (or other meaningful matter) to the context of their use" (Krippendorff, 2004, p. 18). He mentions that according to this definition, "text" is not restricted to written material but also includes, e.g., works of art, images, maps, sounds, signs, and symbols (Krippendorff, 2004). Duriau *et al.* (2007, p. 18) note that "exploratory and interpretive research is more likely to rely on primary data such as interviews, field notes, videotapes, and open-ended questions to surveys". In their meta-analysis, they examine the data sources of the content analysis literature in organization studies. In addition to annual reports, proxy statements, trade magazines, interviews, and open-ended questions in surveys and mission statements, they identify transcribed videotapes as a source (cf. Duriau *et al.*, 2007, p. 16).

In this study, the analysis of the best-practice presentations is based on the videos, the corresponding transcripts, and the presentation slides as three sub-types of sources that were integrated into the data collection and analysis procedures (see Table 34).

Doc. Group	Description of sources	Quantity
Best-practice presentations	• *The video footage of the best-practice presentations at the Innovation Managers Forums in 2013, 2014, and 2015*	*507 minutes*
	• *Corresponding transcripts (document group 1)*	*104 pages*
	• *Corresponding presentation slides (document group 2)*	*680 slides*

Table 34: Results of the data-gathering procedures[67]

All the presentations given at the annual Innovation Managers Forums were video- recorded and prepared for the purpose of this study. The video recordings lasted between 18 and 49 minutes and delivered 507 minutes of raw material. Based upon those recordings, 104 pages (single-spaced) of corresponding transcripts were generated. The process of transcription is described as "the transfer of an audio or video recording into written form. A transcript usually originates from simply typewriting the recorded content"

---

[67] Author's own table.

(Dresing *et al.*, 2015, p. 21). During the transcription process, the guidelines by Dresing *et al.* (2015) were considered. This study follows the "system for simple transcription" in which "paraverbal and non-verbal elements of communication are usually omitted" (Dresing *et al.*, 2015, p. 23) and a "simple transcript allows faster access to the content of the conversation" (Dresing *et al.*, 2015, p. 25). The videos are transcribed using F4 software. Additionally, the corresponding presentation slides were considered (680 presentation slides).

4.4 Data analysis

Following the presentation of the data collection procedures, details of the data analysis procedures are introduced. Section 4.4.1 provides a detailed explanation of how the data were investigated. Here, especially in the illustration of the analysis steps, support is provided by MAXQDA-software and data saturation is considered. Subsequently, the descriptive statistics as first indicators of the achieved outcomes are presented in section 4.4.2.

4.4.1 The analysis process

The analysis process was designed according to the previously mentioned goal and study design, along with the guidelines established by Mayring (2014, p. 80). Therefore, the following steps for the analysis (step A to step D) were planned, defined and executed in this study (see Table 35).

Steps	Description
Step A	*Definition of coding rules*
Step B	*Working through the material, new category formulation, and subsumption*
Step C	*Revision of preliminary categories and abstraction level*
Step D	*Final working through the entire material*

Table 35: Steps of the analysis process for Study B[68]

In step A, the "definition of coding rules" is done by clarifying i) the theme of categories and selection criteria, ii) the level of abstraction, and iii) the unit of analysis as depicted in Table 36. The coding rules determine the identification of all the descriptive elements of organizational support for both firm internal Innovation Contests (theme of categories) and concrete and specific aspects of organizational support (level of abstraction).

---

[68] Author's own table, referencing Mayring (2014, p. 80).

Properties	Definition
Theme of categories and selection criteria	*All the descriptive elements informing critical success factors of organizational support for firm internal Innovation Contests*
Level of abstraction	*Concrete and specific aspects of critical success factors for organizational support*
Unit of analysis	*- Coding: Meaningful words or phrases* *- Context: All the information in one chapter of the presentations* *- Case: Single best-practice presentation*

Table 36: Definition of coding rules[69]

In step B, "working through the material, new category formulation, and subsumption", the meaningful paraphrases were identified, marked and coded as elaborated in the following: "The first time, material fitting the category definition is found, a category has to be constructed […]. The next time a passage fitting the category definition is found it has to be checked, whether it falls under the previous category […]; if not a new category has to be formulated" (Mayring, 2014, p. 81). In this step, the coding system to structure the data is developed and the material is analyzed to identify the categories and main categories. In this case, the categories represent the constituents and the main categories represent the critical success factors (see Table 37) of the organizational support phenomenon in line with the pertinent literature, which distinguishes between manifest and latent contents: "At one level, the manifest content of the text can be captured and revealed in a number of text statistics. At a second level, the researcher is interested in the latent content and deeper meaning embodied in the text, which may require more interpretation" (Duriau *et al.*, 2007, p. 6). In addition, the 'MECE-Rule' is applied that means that the subjacent constituents of a critical success factor are "mutually exclusive", but "collectively exhaustive".

---

[69] Author's own table.

Content-related layers	Technical identifiers
Critical success factors	*Main categories of the coding scheme*
Subjacent constituents	*Categories of the coding scheme*
Facets and representative actions	*Passphrases (direct quotations from the transcripts) and descriptions and explanations (by the author of the present thesis)*

Table 37: Differentiation of layers within the coding system[70]

In step C, "revision of preliminary categories and abstraction level", the identified categories and the codings are checked as an important element of the process: "After working through a good deal of the material [...], no new categories are to be found. This is the momentum for a revision of the whole category system. It has to be checked, if the logic of categories is clear (e.g., no overlaps)" (Mayring, 2014, p. 81). Following Mayring (2014, p. 13) in the sense of classifying step B as a pilot phase, "in qualitative content analysis the category systems are developed inductively out of the concrete material [...] for the specific study. Therefore, those elements have to be pilot tested as well for gaining methodological strength. This is possibly very easy because the textual material can be processed several times." In addition to the category development procedures, the proof of the abstraction level is important to be tested in this step and "if too many categories had been formulated so that a clear picture of the object does not occur, the level of abstraction should be defined more general" (Mayring, 2014, p. 81).

In step D, "final working through the entire material", the full material is analyzed to identify all text passages that refer to the developed categories. In the final coding, "the whole material has to be worked through with the same rules (category definition and level of abstraction)" (Mayring, 2014, p. 83). After finalizing step D, 961 passphrases were found during the initial coding, with 323 marked paraphrases in the transcripts and 638 codings assigned to the presentation slides. The identified passphrases provide general insights into the research question. Additionally, the created categories and main categories offer a more general overview of the areas of actions that must be considered for the implementation of firm internal Innovation Contests (see Appendix D for the coding schemes).

---

[70] Author's own table.

Data saturation is defined as "the point at which no new information or themes are observed in the data" (Guest *et al.*, 2006, p. 59). Guest and colleagues compared various recommendations inform the pertinent literature, summarizing them as follows: "Bertaux (1981) argued that fifteen is the smallest acceptable sample size in qualitative research. [...] Creswell (1998) ranges are a little different. He recommended between five and twenty-five interviews for a phenomenological study and twenty-thirty for a grounded theory study" (Guest *et al.*, 2006, p. 61).

In this study, 18 cases were considered. Some arguments for assuming that the chosen data set is sufficient to answer the research question adequately are derived from the following facts: a) the investigated group of experts and their understanding of the implementation of Innovation Contests within their firm environments is relatively homogeneous and therefore, the consensus between experts in their domain of expertise is high (cf. Guest *et al.*, 2006, p. 74); b) the data analysis procedures imply that the categories were found after analyzing approximately half of the data material; and c) the sampling strategy is in line with previous research. For instance, Guest *et al.* (2006, p. 59) illustrate that in their study, "Saturation occurred within the first twelve interviews, although basic elements for metathemes were present as early as six interviews. Variability within the data followed similar patterns."

As the first result, the descriptive statistics presenting the frequencies of occurrence for the main categories (representing critical success factors) and categories (representing subjacent constituents) are summarized in section 4.4.2.

4.4.2 Descriptive statistics

Generally, descriptive statistics in qualitative research are encouraged, as indicated by the following quotation: "The qualitative content analysis is a mixed methods approach: Assignment of categories to text as qualitative step, working through many text passages and analysis of frequencies of categories as quantitative steps" (Mayring, 2014, p. 10). In line with this suggestion, the following tables and graphics illustrate the distribution of the identified passphrases and their allocation to the categories and main categories. Although this information is useful and aids in understanding the research process and material (Mayring, 2014), it is not highlighted for any statistical interpretations. First, the distribution of codings and allocations to the twelve identified critical success factors representing single areas of actions are illustrated in Table 38.

The distribution of the initial codings shows that at least 20 and up to 176 passphrases are assigned to the main categories. On average, approximately 80 codings were assigned per main category (representing critical success factors), and approximately 46 codings were assigned per category (representing constituents).

A detailed content-related introduction of the identified areas of action in an engaging work environment (separated into critical success factors, constituents, and facets) is not given here; instead, it is included in the subsequent section on the presentation of the results.

Codes	Critical success factors	Freq	%
CSF_1	Strategic Sponsorship	112	11,6 %
CSF_2	Vision and Strategy	101	10,5 %
CSF_3	Visibility	59	6,2 %
CSF_4	Process and Transparency	61	6,3 %
CSF_5	Communication	77	8,0 %
CSF_6	Incentives	68	7,1 %
CSF_7	Staffing and Community	176	18,3 %
CSF_8	Monitoring	67	6,9 %
CSF_9	Campaign Alignment	20	2,1 %
CSF_1	Preparation	35	3,6 %
CSF_1	Execution	126	13,0 %
CSF_1	Transition	59	6,1 %
	Total:	961	100 %

Table 38: Distribution of codings on the level of main categories (tabular overview)[71]

Next, a more nuanced representation considering the critical success factors and their subjacent constituents on a lower level of granularity and the corresponding frequencies of codings is presented in Table 39. Additionally, the percentages representing the distribution of codings for the identified constituents are illustrated.

Regarding the differentiation of categories and main categories, "as a rule of thumb, a set of ten to thirty categories gives a good overview. But sometimes it would be interesting to bring this set of categories into an order by formulation main categories" (Mayring, 2014, p. 81). In this study, this guideline was considered by using 12 main categories and 21 categories.

---

[71] Author's own table.

Codes	Success Factors	Constituents	Freq. / %
CSF_1	Strategic Sponsorship	C1.1: Active Organizational Support	54 / 5,6 %
		C1.2: Positive Attitude	58 / 6,0 %
CSF_2	Vision and Strategy	C2.1: Organizational Alignment	57 / 5,9 %
		C2.2: Trends and Focus	44 / 4,6 %
CSF_3	Visibility	C3.1: Awareness	38 / 4,0 %
		C3.2: Recognition	21 / 2,2 %
CSF_4	Process and Transparency	C4: Process and Transparency	61 / 6,3 %
CSF_5	Communication	C5.1: Communication Activities	48 / 5,0 %
		C5.2: Supplementary Activities	29 / 3,0 %
CSF_6	Incentives	C61: Rewards	31 / 3,2 %
		C6.2: Gamification	37 / 3,9 %
CSF_7	Staffing and Community Building	C7.1: Staffing	59 /6,1 %
		C7.2: Community Building	117 /12,2 %
CSF_8	Monitoring	C8.1: Progress Monitoring	33 / 3,4 %
		C8.2: Benefits Monitoring	34 / 3,5 %
CSF_9	Campaign Alignment	C9: Campaign Alignment	20 / 2,1 %
CSF_10	Preparation	C10: Preparation	35 / 3,6 %
CSF_11	Execution	C11.1: Ideation	58 / 6,0 %
		C11.2: Discussion	36 / 3,7 %
		C11.3: Evaluation	32 / 3,3 %
CSF_12	Transition	C12: Transition	59 / 6,1 %
		Total:	961 / 100%

Table 39: Distribution of codings on the category level (tabular overview)[72]

---

[72] Author's own table.

4.5   Presentation of results

This section presents the findings of the study based on the qualitative content analysis of best-practice presentations. In line with Mayring (2014, p. 13) "at the end of processing the study it is important [...] to present the results in a broad descriptive sense and in the more specific sense of answering the research question". Accordingly, both the outcomes in the form of the identified critical success factors, underlying constituents, and facets of an engaging work environment and a framework for clustering the critical success factors is introduced in section 4.5.1. Next, details regarding the single success factors and subjacent constituents are given by presenting the identified facets and representative actions (sections 4.5.2 through 4.5.4). The chapter ends with a summary of Study B (section 0).

4.5.1   A framework of critical success factors for Innovation Contests

The aim of this research is to identify important determinants of an engaging work environment for the organizational support of Innovation Contests embedded in an organizational environment. The identified critical success factors, constituents and facets represent explanations of the innovation experts and their information sharing during the best-practice presentations and cover a broad range of aspects that must be considered during the implementation of firm internal Innovation Contests as complex undertakings.

On a higher level, this study's findings result in a framework that summarizes success factors on three differentiated levels (see later in this section) for the sake of both clarity and structuring the outcomes. With respect to the sense and purpose of conceptual frameworks, this study follows the expressions by Elaine Botha (1989, p. 51), who explains that "conceptual models are used to provide a perspective or focus in order to interpret phenomena [...]. These conceptual models form the basis of all theorizing". The need for holistic theories is further highlighted so that "one characteristic distinguishes theoretical concepts and theoretical knowledge from everyday knowledge: The fact that it is of an abstract nature in contrast to our everyday knowledge which deals with concrete states of affairs and events. This characteristic is simultaneously the strength of all theorizing" (Elaine Botha, 1989, p. 54). In their work highlighting the determinants of organizational innovation, Crossan and Apaydin (2010, p. 1169) state their aim as follows: "We propose a comprehensive framework of organizational innovation which provides an overarching structure than can link different theoretical units into a coherent whole". In a similar manner, the decision to use a three-layer structure to represent the critical success factors turned out as detailed below.

In addition to the endorsement for the use of a framework in general as described above, the reasons for differentiating several layers in this study could be found in the literature (Crossan and Apaydin, 2010; Amabile and Gryskiewicz, 1989). In their multi-dimensional framework of organizational innovation, Crossan and Apaydin (2010) differentiate between management on the strategic level, the organizational level and the process level when explaining the determinants of innovation. Explaining the influence of the strategic level, they note that "encouragement should come, ideally, from the highest levels: The chairman, president, or CEO" (Amabile and Gryskiewicz, 1989, p. 233). Next, the influence of support on the underlying organizational level is highlighted: "The motivation to innovate can, however, also be important. These levels of management are often responsible for communicating and interpreting the orientation of those at the highest levels" (Amabile and Gryskiewicz, 1989, p. 233). Finally, the influence on the concrete project or process level is highlighted. Crossan and Apaydin (2010) focus on the process level for supporting innovation. Here, a business process is a meta-construct that "studies how organizational processes convert inputs into outputs" (Crossan and Apaydin, 2010, p. 1173). In addition, Cooper and Kleinschmidt (2007, pp. 53 ff.) note that "success at the company or business level may differ from success at the project level" and furthermore, "perhaps the most fundamental flaw is the fact that research done at the project level often fails to identify those company practices that decide success".

Consequently, three layers are distinguished in this study (see Table 40). First, the "strategic level" represents all the critical success factors linking the innovation initiative with the organization and offering both support and a strategic orientation. Second, the "initiative level" considers all areas of actions that are important for the entire innovation initiative, meaning the sum of all contests and accompanying measures. Third, the "contest level" focuses explicitly on critical success factors on the level of Innovation Contests. The logic of the framework and relationships between the levels follows the rule that critical success factors on a lower level require the factors that are represented on a higher level. For instance, the engaging work environments aspects on the initiative level require the consideration of the identified areas of actions described at the strategic level. Conversely, the factors on the higher level incorporate the factors mentioned on the lower levels.

Designation	Description
Strategic level	*All the critical success factors that are relevant to linking the Innovation Initiative and the organization*
Initiative level	*All the critical success factors with an innovation initiative wide relevance*
Contest level	*All the critical success factors with an Innovation Contest-wide relevance*

Table 40: Levels of the conceptual framework[73]

Additionally, the differentiation of three levels has turned out to be suitable for the data analysis procedures and the description of the outcomes. Moreover, differentiation can be strengthened by explanations in the researched data material, as illustrated, e.g., by the following quote:

*"We **have three levels of sponsorship in place.** There is a C-Suite sponsorship on the strategic innovation areas and that's really allowing or showing people that it's ok to participate in the program. Our top management is behind the program."*

- Innovation Program Manager, Case L

The following chart illustrates the framework that consolidates the 12 identified success factors driving an engaging work environment and organizational support for Innovation Contests in a corporate environment (see Figure 23).

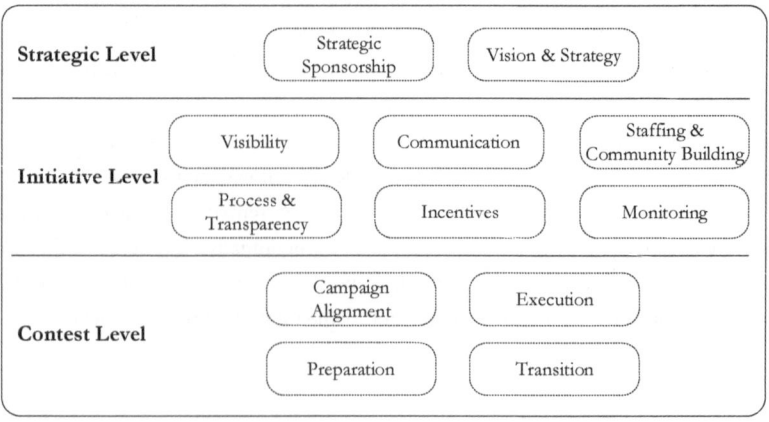

Figure 23: The critical success factors of firm internal Innovation Contests[74]

---

[73] Author's own table.

First, on the strategic level, the framework encompasses two areas of actions representing the first critical success factors, namely, a **Strategic Sponsorship** and the definition of a **Vision and Strategy**. Here, upper management is especially involved in strategically leading the initiative. At the initiative level, there are six areas of actions, including **Visibility**, **Process and Transparency, Communication, Incentives, Staffing & Community Building**, and **Monitoring**. On that level, the managers of the innovation initiatives are especially responsible for the design and adherence of these important success factors that are critical to the creation of an engaging work environment and associated aspects of such a project. On the contest level, the areas of actions must be considered separately for each contest execution. Here, the **Campaign Alignment, Preparation, Execution**, and **Transition** are the identified critical success factors of an engaging work environment. Subsequently, the decompositions of areas of actions on the differentiated levels, namely, on the strategic level (section 4.5.2), the initiative level (section 4.5.3) and the contest level (section 4.5.4), are presented in detail. Therefore, the following presentations of the areas of action consistently follow the identical structure. Followed by the introduction of the critical success factor, for every subjacent constituent, an introduction and a reference to the state of the art in the pertinent literature are provided to increase the connection with former research. Next, an overview of the identified facets and representative actions is provided. Finally, direct quotations from the presentation transcripts are highlighted to offer additional insights, enhance the meanings of the facets, and increase the transparency of the research process.

4.5.2   Decomposition of support factors on the strategic level
On the top level, the first two areas of actions of an engaging work environment for a support of firm internal Innovation Contests are differentiated into **Strategic Sponsorship** and **Vision and Strategy**. Both are presented below.

4.5.2.1   Strategic Sponsorship
This first critical success factor consolidates all actions of the top management level regarding the **Strategic Sponsorship** of the entire innovation initiative. Sponsorship is important for crowd-based innovation activities, as highlighted in the pertinent literature: "The interaction between the sponsor and solvers is an important mechanism in crowdsourcing. Lack of interactions can create a negative word-of-mouth message about

---

[74] Author's own figure; a former version was discussed at the Open and User Innovation Society Meeting, see Hoeber (2015).

the sponsor company" (Zheng *et al.*, 2014, p. 216). Two subjacent constituents to establish a strategic sponsorship for the innovation initiative are revealed through this study, namely, **Active Organizational Support** and a **Positive Attitude** about the innovation initiative.

First, **Active Organizational Support**, e.g., through proactive participation, visible support, involvement and active steering by upper management helps increase the seriousness and importance of the entire Innovation Contest initiative and all single Innovation Contest executions. Additionally, support by executive management helps in changing the mindset of the workforce and increase the encouragement of employees to spend time in the program as part of their daily work. Additionally, it is noted that there is sufficient mandate and responsibility that the best ideas can be implemented after their selection.

The pertinent literature complains that "much of the early-stage front-end process is not well understood or managed at the organizational level" (Reid and De Brentani, 2004, p. 175). Furthermore, because of "the lack of involvement or understanding by upper management (where strategic, structural, and resource planning occurs), the process may come to a near standstill" (Reid and De Brentani, 2004, p. 175). Management support, involvement and commitment have been identified as important factors in innovation activities (Govindarajan, 2011; Kuratko *et al.*, 1990; Cooper and Kleinschmidt, 1995; Jarvenpaa and Ives, 1991). Management support is also highlighted by Stol and Fitzgerald (2015). They argue, "perhaps one of the most important factors in starting successful inner-source initiatives is top-level management support" in which "management made a strong commitment" because contributors "must be motivated to voluntarily participate in this model to make it successful" (Stol and Fitzgerald, 2015, p. 66). Stol and his colleague highlight a good example in which the investigated firm "has encouraged its business units to use, and actively engage in, inner source" (Stol and Fitzgerald, 2015, p. 66).

Table 41 illustrates the identified facets and representative actions of **Active Organizational Support** as the first constituent of the critical success factor **Strategic Sponsorship**.

Facets	Representative actions
Encouragement of employees	• *Encouragement of employees both to join the program and to participate (e.g., via messages and invitations)*
Involvement and Participation of Sponsors	• *Involvement, commitment, and attitude of strategic sponsor(s), top and upper management; active participation, interaction, and attention in the initiative by managerial colleagues*
Active steering of the Innovation Initiative	• *Increased visibility and active steering by the management level for both the entire initiative and Innovation Contests*
Sponsorship on various levels	• *Establishment and anchoring of the strategic, tactical, and crowd sponsorship, along with supervisor support and encouragement (e.g., permission to attend, encouraging, free space)*

Table 41: *Active Organizational Support* as a constituent of *Strategic Sponsorship*[75]

The following quotes from the innovation experts underline the importance of **Active Organizational Support** for engaging employees as voluntary participants in firm internal Innovation Contests. First, the facet of an *Encouragement of Employees* to participate in the innovation initiative is highlighted. It can be described as a corporate mandate clarifying that the initiative is supported by upper management. One quotation provides more information:

> *"Finally, we had a good success to get a paper from the board also the Executive board of […] where all members signed here. A letter again, where also **the management said please join the program and you will have time for it**."*

> - Expert, Case O

In addition, the *Involvement and Participation of Sponsors* is an important facet. For instance, one innovation expert stated the following:

> *"What we are really looking for there is **sponsorship from someone in the client's business**, someone who wants us to help them address a business challenge, a problem."*

> - Expert, Case R

---

[75] Author's own table.

Additionally, the data analysis reveals the facet of *Active Steering of the Innovation Initiative*. Following is one example from the transcripts:

> *"On the other hand, we need the sponsors and advocates at executive level to give the clear signal down to the middle management [...]: 'Yes, we want this. We want this change, and we want to give this little plan a chance to survive.'"*

\- Expert, Case I

Finally, *Sponsorship on different levels* could be revealed as important. One quotation provides an example:

> *"We have three levels of sponsorship in place. There is a C-Suite sponsorship on the strategic innovation areas and that's really allowing or showing people that it's OK to participate in the program.* **Our top management is behind the program.**[76]*"*

\- Expert, Case L

Second, the critical success factor Strategic Sponsorship includes the establishment of a **Positive Attitude** about the innovation initiative by clarifying the important role of innovation and increasing organizational confidence and shared values. Additionally, by acting as inspirational leaders, enthusiasm and spirit, applied change management, a positive attitude towards cultural change and a common philosophy could be solidified.

According to the pertinent literature, "another important factor which enables innovation as a process is organizational culture" (Crossan and Apaydin, 2010, p. 1172). Leadership is especially important as a determinant of the perception of the climate for innovation and have an impact on innovative behavior; individual innovation in the workplace is highlighted in the literature (Scott and Bruce, 1994). The identified facets and representative actions of a **Positive Attitude** about the innovation initiative as constituent of **Strategic Sponsorship** are illustrated in Table 42.

---

[76] This quotation was mentioned in section 4.5.1, which introduces the framework.

Facets	Representative actions
Employees' Attitude towards Change	• *A positive attitude, enthusiasm, and inspiration of employees with respect to cultural change (mindset and failure tolerance, international varieties in culture and skill sets) and a common philosophy; inspirational leadership both to catalyze change and to increase enthusiasm and spirit*
Role of the Innovation Initiative	• *Establishing innovation as a fundamental role and core competence of the firm, corresponding the establishment of the innovation initiative as an integral aspect of the generation of novel ideas*
Organizational Confidence in the Innovation Initiative	• *Creation of organizational confidence, beliefs and shared values regarding the innovation initiative, the avoidance of distrust, criticism and resistance to the innovation initiative*

Table 42: *A Positive Attitude* as a constituent of *Strategic Sponsorship*[77]

Below, quotations that exemplify a **Positive Attitude** about innovation initiative are provided by the innovation experts. First, *Employees' Attitude towards Change* is highlighted as an important facet. For example, the innovation experts stated as follows:

"We have to engage the innovators, we have not to force the ideas but engage the innovators to find them so **that we can really create an innovation culture, a protected space** *for the 'crazy' people."*

– Expert, Case I

***

"It is different from how we work normally [...] and **this is a big cultural change to dare to go out of your daily business."**

– Expert, Case I

***

---

[77] Author's own table.

"Finally, **they voluntarily engage in the activity. It's not a forced thing.** *Again, I can give another client where they actually required people to hit certain thresholds of activity [...].* **A lot of people get frustrated."**

- Expert, Case A

In addition, both the *Role of the Innovation initiative* and *Organizational Confidence in the Innovation Initiative* are revealed as facets of a *Positive Attitude* during the data analysis. One innovation expert provided the following explanation:

"It's a bunch of cultural topics playing a role: 'Not invented here', I know my business, who should know it better? We have many ideas but no time to do/to run them. We don't need additional ideas. These sentences are coming in."

- Expert, Case K

Below, **Vision and Strategy** as an additional area of action is presented.

4.5.2.2 Vision and Strategy

The success factor *Vision and Strategy* pays attention to the alignment between the initiative, the strategic goals and other areas of the organization, summarized in this study in the constituent of **Organizational Alignment.** Second, the anticipation of trends and business opportunities and the definition of a focus for the initiative are consolidated in the constituent of **Trends and Focus**.

The element **Organizational Alignment** ensures matching and linking the goals of the initiative and each Innovation Contest in line with both organizational strategy and innovation goals. Furthermore, a strategic alignment helps initiatives fulfill the needs of managers and the entire organization and can solve existing business challenges. Additionally, the entire innovation initiative needs to be aligned to other activities within the firm, especially other innovation initiatives.

Regarding the state of the art in scientific literature, "an explicit innovation strategy [...] is a primary managerial lever and helps to match innovation goals with the strategic objectives of the firm" (Crossan and Apaydin, 2010, p. 1172). Additionally, the literature has highlighted the importance of the new product strategy for innovations: "A clear and well-communicated new product strategy for the company, that is: There were goals or objectives for the company's new product program [...]; The role of new products in achieving company goals was clearly communicated to all in the firm; there were clearly defined areas—specified areas of strategic focus" (Cooper and Kleinschmidt, 1995, p. 384).

Table 43 illustrates the identified facets and representative actions of an **Organizational Alignment** as a constituent of the critical success factor **Vision and Strategy**.

Facets	Representative actions
Strategic Alignment of the Initiative	• *Matching, linking and strategically aligning the initiative with corporate strategy, innovation strategy and goals*
Strategic Coordination	• *Strategically coordinating activities and measures regarding the initiative within the organization, between business units and requesters, corporate functions and other R&D activities*
Shared Vision and Goal Setting	• *Developing and composing a shared vision and goal setting for the entire Innovation Initiative in accordance with the environment and the organization*

Table 43: *Organizational Alignment* as a constituent of *Vision and Strategy*[78]

Excerpts about the need for **Organizational Alignment** are found in various passages mentioned by the innovation experts when discussing best practices. *Strategic Alignment of the Initiative* and *Strategic Coordination* within the organization are the highlighted elements:

"What we would like to address to the management was first of all the link of our program to the corporate strategy. Then, the relationship to the identified megatrends and our search fields. Then, the innovation process itself and surely also the Portal and Tour."

- Expert, Case O

***

"We don't really know what the organization cares about [...]. Well, let's try and get better aligned. Let's try and think much more carefully about where the organization is heading, what we care about and make sure that our cognitive innovation programs are completely aligned with the strategic innovations of the organization."

- Expert, Case A

Additionally, *Shared Vision and Goal Setting* is identified as an important facet, as highlighted by the following quote:

---

[78] Author's own table.

"You have people here that know the market and are able to create a vision. You have here in the company different people that know how to operate our work facilities and how best to apply innovation."

- Expert, Case J

Second, the constituent **Trends and Focus** stands both for the identification, collection and administration of trends and for the definition of the focus and appropriate direction of all the initiative's innovation activities. All the information should be easily accessible to all employees and managed in a centralized trend pool.

The pertinent literature has highlighted the need to scan for trends, technology, market changes and search fields for new business creation as an opportunity for innovation development in organizations: "The purpose of scanning is therefore to identify key events and trends, and consequently to envisage the possible impact on the existing order, and the opportunities for innovation" (Tang, 1998, p. 301).

Table 44 gives a representation of the identified facets for **Trends and Focus** as the second constituent of the success factor **Vision and Strategy**.

Facets	Representative actions
Trend Collection and Administration	• *Identification, collection and management of trends, business and market opportunities, and search fields; managing upcoming trends, forecasts, business potentials and market opportunities, and new technologies in a central trend pool*
Focus of the Initiative	• *Focus on defined areas of innovation within the initiative with an appropriate direction of activities and future scenarios as part of the innovation strategy*

Table 44: *Trends and Focus as a constituent of Vision and Strategy*[79]

In this study, *Trend Collection and Administration* is revealed as the first important facet:

"There is trend scouting going on (for the market pull part) within our company at the marketing department. So they are really focusing on what changes are going on in the market, what new markets are emerging, and they put it in a database in the software; and then there is an evaluation going on."

- Expert, Case N

---

[79] Author's own table.

***

*"**We often face completely new business fields** where we are not experts in them. Should we go this way and go into, let's say, a red ocean, maybe. The implementation and commercialization of solutions in completely new business fields is a challenge."*

- Expert, Case K

***

*"**It will be a trend pool for the technology and innovation management** and the top management. We will also use this opportunity in the future to collect trends from our employees."*

- Expert, Case N

In addition, based on the above-mentioned facets, it is important to define the focus of the entire innovation initiative as consolidated in the next formulated facet, *Focus of the Initiative*. The innovation experts made the following statements:

*"I think this is the core idea of why I'm bringing up these key opportunity areas. **You need to focus energy somewhere.** You cannot spread it out and think about innovation [...]. Let's come up with a good idea and see if the firm likes it, **because you get 99 % from outside of where you would like to go."***

- Expert, Case Q

***

"The first thing is when we talk about market pull and technology push. These are the basic approaches when we talk about innovation management. The question for us was do we focus on the one part or on the other part. At the beginning we said ok, the focus is on the customer needs."

- Expert, Case N

Subsequent to the presentation of **Strategic Sponsorship** and **Vision and Strategy** as the first two identified areas of action, additional areas that belong to the initiative level are presented below.

4.5.3 The decomposition of support factors on the initiative level

Most of the success factors of organizational support for Innovation Contests are identified on the initiative level, highlighting initiative-wide relevance. The identified success factors are **Visibility, Process and Transparency, Communication, Incentives, Staffing and Community Building,** and **Monitoring**. All the critical success factors represented separate areas of action and are usually not considered separately for each single

Innovation Contest execution, but for the over-arching innovation initiative. In subsequent sections, each factor is presented in more detail.

4.5.3.1 Visibility

The critical success factor **Visibility** focuses on the need for the entire initiative to maintain high visibility to all interested employees.

The literature has mentioned visibility as important driver. For instance, Wooten and Ulrich (2015) provide evidence of visibility in Innovation Contests. The researchers differentiated two modes (blind and unblind) of publishing ideas in a contest. The results indicate that the visibility of entries "generates more entries by increasing the number of participants" (Wooten and Ulrich, 2015, p. 1). **Visibility** as a critical success factor indicates active support for an increase in both the **Awareness** and **Recognition** of the entire innovation initiative based on Schmidt *et al.*'s (2002, p. iii) statement about computer-supported cooperative work-systems "despite the growing interest in awareness, and the recognition that it is of critical importance to the successful development of systems to support cooperative activity".

First, regarding **Awareness** of the initiative, it is important to conduct an active promotion regarding, *inter alia*, the visibility of the innovation initiative and content (both online and offline). For instance, useful and valuable content and contributions should be spread over several channels in the company to make it available for all employees so that these contributions can serve as either inspiration or motivation. This is especially important because undecided employees might be hesitant and wait to see what happens to other contributions before they will join the initiative. Additionally, customizing the platform in a manner that considers corporate identity guidelines is important for employees' attention and awareness and finally, their intention to participate.

An overview of the identified facets and representative actions of **Awareness** as a constituent of the critical success factor **Visibility** is presented in Table 45.

Facets	Representative actions
Promotion of the Initiative	• *Promotion of the innovation initiative to increase visibility, attention, and awareness of the approach within the company (e.g., posters, tours, letters, Post-Its, media channels)*
Promotion of Content and Innovators	• *Promotion of the content (contributions and outcomes, e.g., solutions) and innovators, making it available for all employees and publishing inspirations and success stories to increase awareness*
Branding and Customizing	• *Branding and customizing of the IT platform based on corporate identity guidelines (including key visual and slogan)*

Table 45: *Awareness* as a constituent of *Visibility*[80]

For the *Promotion of the Initiative* and for the *Promotion of Content and Innovators,* excerpts from the innovation experts can be cited, including the following passages:

"It was fun to observe this, and the employees[81] **really started discussions like** *'what is this?' and 'have you heard about this?' and 'What are they doing?' and so on. This was really* **something which gives some awareness to the people."**

- Expert, Case O

***

"And for us in innovation that's a challenge about how do we **attract the attention of those people** *that have a very clear imperative to actually keep the lights on, do the day job."*

- Expert, Case R

In addition, *Branding and Customizing* of the IT platform is highlighted. One example is as follows:

---

[80] Author's own table.
[81] For the sake of clarity, the author replaced "they" with "the employees".

"You all know this platform with this standard look. We said that we need something in corporate colors, so we redesigned it. Today it looks like that, so we have a very light version with a picture showing our search fields. [...] So it's a design to create recognition."

- Expert, Case O

Second, to increase the **Recognition** of the initiative, the spreading of user-generated values and success via several channels in the company, *inter alia*, should be considered. For instance, the sharing of implemented ideas or success stories can show how the platform can help businesses units with their challenges and business problems. Publicly available content might serve as inspiration or motivation for undecided colleagues. In addition, recognition could be achieved not only through constant awareness, sustainability and belief but also by the acknowledgement of contributions and process flexibility.

The identified facets and representative actions of **Recognition** of the entire innovation initiative as a constituent of the critical success factor **Visibility** are illustrated in Table 46.

Facets	Representative actions
Sharing of Results	• *Achievement of recognition by sharing the achieved results*
Sharing of generated Values and Success	• *Achievement of recognition by clarifying and publishing the company's successful projects, outcomes and business values*

Table 46: *Recognition* as a constituent of *Visibility*[82]

The following quotes illustrate the necessity of the recognition facet. First, a broad *Sharing of Results* is mentioned:

*"**We really make sure that the results are shared** on our corporate intranet and through corporate communication materials. We also interviewed winners of campaigns to have them let to say in how they experience the process."*

– Expert, Case L

∗∗∗

---

[82] Author's own table.

"This also served another goal. This is showing results. So **by showing these good results, we get other departments that get interested in the program** *and want to participate and want to organize their own campaign."*

– Expert, Case L

Finally, by communicating and a *Sharing of generated Values and Success*, recognition of the employees for participating in the innovation initiative can be enhanced. The following excerpt underlines this effect:

"We really see that those spikes in between really follow communication moments. That's really my big advice to keep communicating and keep showing the results of the campaign. **At the end of the campaign, we have caused recognition of our employees."**

– Expert, Case L

Subsequently, details on **Process and Transparency** as additional areas of action are provided.

4.5.3.2 Process and Transparency

On the initiative level, the critical success factor of **Process and Transparency** consolidates all aspects of an engaging work environment related to the clarity of the process and procedures during the execution of Innovation Contests within the Initiative. This study highlights that potential contributors want to understand, e.g., how the initiative proceeds, what they need to do, and what they can do during the various activities and steps.

In the scientific literature, transparency within a group of interacting people is regarded, for instance, "as the degree to which the communication network is sufficiently clear and accessible, in order to let everyone understand the inputs and progress made (Hamel, 1991). Limited transparency implies that members of a network have problems identifying the relevant persons to transfer information to or to obtain information from" (Moenaert *et al.*, 2000, p. 364). Process transparency is also evaluated as important to inner-source open-source projects (cf. Stol and Fitzgerald, 2015). Here, the process must be transparent and "generally open to anyone who wishes to be involved" and furthermore, "without such transparency, developers won't be able to 'lurk' and will have no way to contribute" (Stol and Fitzgerald, 2015, p. 65). Additionally, the clarification of roles and responsibilities is highlighted. "Nevertheless, it might be important to formalize a number of roles in the core team that is managing an inner-source project. Which roles are necessary will depend on the organization's context, and they might emerge as needed" (Stol and Fitzgerald, 2015, p. 65). In particular, the participants must obtain a suitable illustration and understanding of the process and procedures, the various steps, responsibilities and desired behavior and the evaluation criteria. For instance, most of the participants want to know what happened to their contributions during and after the contest. Practical experience shows that training for potential contributors is important both to increase understanding of the procedures and inner workings and to dismantle doubts and worries.

The facets and representative actions of the critical success factor of **Process and Transparency** are illustrated in Table 47.

Facets	Representative actions
Transparency of the Process	• *Improve the transparency, clarification, and understanding of the process and variations through explanation, illustration and communication of the process and single steps, specialties, and guidelines, e.g., by offering information possibilities and training for employees*
Clarification of Roles and Responsibilities	• *Clarifying roles and responsibilities as the basis of an efficiently execution of the process and creating trust*
Process Design and Flexibility	• *Ensuring a flexible process design through the definition of a complete, customized and flexible process, e.g., as a possibility to react to variations and changing requirements*

Table 47: Representative actions for the critical success factor *Process and Transparency*[83]

The innovation experts made the following three recommendations related to the *Transparency of the Process*:

> *"People were sharing content but weren't aware of what was going to happen next. The program essentially was closed. Once you submitted your content, it was evaluated. You had no idea what the next step was, what the implementation path would be, whether your idea was even taken forward or not. Of course, over time, that degraded the confidence the organization had in the process as a whole."*

- Expert, Case A

***

> *"We also have engineers we need to talk to; that needs a structured process. We have gates. We have steps. And again, it's kind of a structured way to express a message."*

- Expert, Case Q

---

[83] Author's own table.

***

"It is a project pool for top management, for the managers, also for the project managers, **to keep everything transparent what we are doing within the innovation management."**

- Expert, Case N

Additionally, the *Clarification of Roles and Responsibilities* could be revealed as an important facet. One example is as follows:

"Mostly it is the lack of clarity, what is our motivation; [...] because it is a hype? Why **should we do this? Who runs it, what are the action points? Who is responsible for what?"**

, - Expert, Case K

In addition, the opportunities provided by high *Process Design and Flexibility* are highlighted. The following quote underlines their importance:

"We really have found the software for us and we are really excited about it because **it has such a lot of opportunities to be able to implement different processes,** *different workflows in parallel [...]. We already have different objects in the system and different evaluation workflows, on the trend level, on the search field level, and then at the end on the idea level."*

- Expert, Case N

Next, I highlight the critical success factor **Communication** as an additional area of action for an engaging work environment that supports the success of firm internal Innovation Contests.

4.5.3.3 Communication

Communication is a central and critical factor in the success of an engaging work environment that functions as an organizational support both for the innovation initiative and for running Innovation Contests. The critical success factor **Communication** could be differentiated into the subjacent constituents **Communication Activities** and **Supplementary Activities**, with the former targeting the core communication activities related to the execution of innovation campaigns and the latter summarizing supplementary activities, for instance, considering offline or marketing activities.

The positive influence of intra-firm communication has been mentioned in the pertinent literature (e.g. Bolton and Dewatripont, 1994). In addition, an investigation of intra-organizational communication, encouragement, participative climate and interaction as determinants of organizational innovation is provided by Kivimaeki *et al.* (2000). They highlight the importance of communications for innovation activities: "The flow of information inside the organization also plays a critical role in promoting innovations. A high level of internal communication may contribute to innovation at several points of the development process" (Kivimaeki *et al.*, 2000, p. 34). In addition, the role of managers is highlighted because "the successful adoption of innovations may depend largely on the support and co-ordination during the innovation process that only managers can provide" (Kivimaeki *et al.*, 2000, p. 34). Management communication can contribute to innovation through fluent internal communication for various purposes, e.g., problem-solving, cross-fertilizing, and implementing ideas (Kivimaeki *et al.*, 2000). Management's role in internal communication is important because the success of innovation largely depends on the support and co-ordination of managers (Kivimaeki *et al.*, 2000; Scott and Bruce, 1994).

First, the constituent **Communication Activities** accompanies rigorous communications, the clarification of all aspects of the platform, announcements of new activities, ongoing communications during campaigns and establishing the platform as single point of access. The analysis shows that an accruing lack of communication is very harmful for such innovation initiatives and therefore, regular communication is of high importance. In addition, regular feedback for contributions in running Innovation Contests should be highlighted.

An overview of the identified facets and representative actions for **Communication Activities** as a constituent of the success factor **Communication** is given in Table 48.

Facets	Representative actions
Rigorous Communication	• *Rigorous communication based on a communication strategy and guidelines (incorporating corporate communications, newsletters, intranet) to increase credibility and belief in the initiative*
Announcement of New Activities	• *Announcement of new activities, especially new campaigns and new phases via several channels*
Ongoing Communication and Feedback	• *Ongoing communication and Innovation Contest support with feedback on the status of contests, contributions, and action points*
Single Point of Access for Innovation Activities	• *Establishment of the platform as a single point of access for employees and innovation-related activities*

Table 48: *Communication Activities* as a constituent of *Communication*[84]

First, excerpts for the facets of *Rigorous Communication* and the *Announcement of new Activities* can be found in several passages, for instance:

*"**We need a communication strategy, also internal,** because we need to increase the awareness for this activities, also insight the companies. We need to win the promotors inside the company. We need to get the sponsors who are by the way, giving us the budget."*

- Expert, Case K

***

*"**Use different methods to communicate because different employees like different methods,** and I also think you should have multiple touch points to really get the employees[85] going. But we can, of course, do it even more differently, so that's why we reached out."*

- Expert, Case L

***

---

[84] Author's own table.
[85] For the sake of clarity, the author replaced "them" with "the employees".

"Communication is really important. We really see that **as long as we keep communicating about the program, participation is high** *and people are participating."*

- Expert, Case L

In addition, *Ongoing Communication and Feedback* is highlighted in this study as follows:

"Another issue we found is communications. **Feedback was not great** *So some of the people that had stuck their necks out, that had said, 'Yeah, I'm prepared to participate in this', hadn't heard what happened next,* **so they stepped away, they became disengaged."**

- Expert, Case A

In addition, the platform as a *Single Point of Access for Innovation activities* is highlighted. One example is as follows:

"The philosophy of our company by using the software as an innovation management platform is like I mentioned at the beginning, **the single point of access for the employees.** *I talked about innovative ideas: This is something our people are able to submit on the platform."*

- Expert, Case N

Next, **Supplementary Activities** as a constituent of **Communication** merges offline and marketing activities to support the entire process, and establishes contact points and marketing activities, road shows, newsletters, and information at the intranet to help increase the up-to-date knowledge of the workforce. These help create excitement and enthusiasm for the entire initiative through both offline and marketing activities. Occasionally, new or not previously interested employees want to join the program. Therefore, it is important to keep them up to date and to provide information about the initiative.

In the literature, Jung *et al.* (2012), for instance, have argued that interactions, collaborations and discussions in enterprise-wide innovation challenges can be supported by a combination of online and offline activities. Adamczyk *et al.* (2010, p. 3) have highlighted that "concerning media, Innovation Contests can be run online, offline or mixed mode".

The identified facets and representative actions for **Supplementary Activities** as a constituent of the success factor **Communication** are given in Table 49.

Facets	Representative actions
Supplementary Offline Activities	• *Organization of supplementary activities, e.g., for process support, workshops before, during and after the contest execution, invitation to work on ideas in inspiration spaces, etc.; establishment of well-known and available contact points, e.g., open space to innovate in a practical manner and to meet people from various horizons*
Marketing Activities	• *Organization of marketing activities (e.g., fairs)*

Table 49: *Supplementary Activities* as a constituent of *Communication*[86]

Regarding the facet of *Supplementary Offline Activities*, first the combination of using the software and presence workshop is mentioned. This study observes the following quotes:

"**We use both the software and face-to-face workshops** *to support a wide range of innovation styles.*"

- Expert, Case R

***

"*So we started, again, to do normal workshops to bring people to ideas, to inspire people to create ideas.* **If we gather, some ideas in those workshops, we put them into the system.**"

- Expert, Case P

***

"*So we use the software in face to face workshops [...]. They were interested in us in being able to do that in another way of* **being able to use the software**[87] **in an accelerated workshop environment.**"

- Expert, Case R

In addition, *Marketing Activities* are seen as important:

"*Well,* **stuff like being offline available to really**, *hey guy, please join and start engaging. And we see on a local level often that things [...]* **and then the ideas were submitted.**"

- Expert, Case L

---

[86] Author's own table.
[87] For the sake of clarity, the author replaced "it" with "the software".

In the following section, **Incentives** as an additional area of action in an engaging work environment supporting firm internal Innovation Contests are presented.

4.5.3.4 Incentives

In the state-of-the-art scientific literature, **Incentives** are a frequently highlighted element of Innovation Contests and a central topic of investigations in several publications in the area (cf. Jeppesen and Frederiksen, 2006; Boudreau *et al.*, 2011; Hutter *et al.*, 2011; Cahalane *et al.*, 2013). Adamczyk *et al.* (2010, p. 4) illustrate that "motivation can be induced via extrinsic motivators (awards and prizes), intrinsic motivators (reputation in the relevant community) or mixed mode". Piller *et al.* (2012) differentiate between economic exchanges (e.g., monetary exchanges) and social exchange (e.g., fun, task achievement). Accordingly, the critical success factor of **Incentives** consolidates areas of action in two constituents, in concrete **Rewards** (including, *inter alia*, a differentiation of rewards) and **Gamification** (including elements of reputation, status management, and gamification features) as presented and detailed below.

First, by offering monetary and non-monetary **Rewards** as an integral part of Innovation Contests, the participants' various objectives and motives could be taken into consideration.

According to one confirmation from the pertinent literature, "the motivation to innovate can be increased by adequate incentives" (Schmelter *et al.*, 2010, p. 165). As one important differentiation, in addition to offering rewards for the generation of novel ideas, others who helped during the elaboration and improvement of ideas, e.g., by showing desired behavior such as supporting or cooperating, should be incentivized. Adamczyk *et al.* (2010, pp. 3 f.) mention that "to foster participation, the organizer establishes a reward system to motivate the participation of the target group—adapted to its needs".

An overview of the facets and representative actions for **Rewards** as a constituent of the success factor **Incentives** is provided in Table 50.

Facets	Representative actions
Reward Differentiation	• *Offering monetary and non-monetary rewards (e.g., money, awards, prizes, partnering, public idea presentation) to consider the participants' various motives*
Appreciation of Contributions	• *Appreciation of contributions through the recognition, acknowledgement, consideration, and recognition of innovative ideas and contributions (e.g., comments, voting)*
Appreciation of Participants	• *Appreciation of participants in the innovation initiative through the recognition of participants, their roles, desired behavior and supportive actions*

Table 50: *Rewards* as a constituent of *Incentives*[88]

The following quotes illustrate the importance of **Rewards**, with an adequate *Reward Differentiation* as the first facet:

*"**You reward people who get good quality ideas into the system** at a much higher level, and those that get their ideas implemented at an even higher level. So we are always driving people towards quality, **but we still recognize the good behaviors we need in order to get us there.**"*

- Expert, Case A

The *Appreciation of Contributions* is revealed as a second facet, as highlighted in the following quotation:

*"**Recognize quality contributions**, we talked a lot about this. View the campaign ideas and comments. Highlight these in your communications. People take their cue by what is recognized."*

- Expert, Case A

Additionally, the *Appreciation of Participants* is highlighted:

*"**It's about trying to have some incentives** and support for the local views, our local team to engage into innovation activities; **also about recognizing and rewarding people** and the teams that are engaging in innovation."*

- Expert, Case J

---

[88] Author's own table.

Next, **Gamification** is revealed as the second constituent of the critical success factor **Incentives**. For instance Kubátová (2012) highlight gamification and reputation features and their importance to cooperation support in online communities in a case study as follows: "The company created an on-line community with gamification, reputation features, and awards that engage and motivate" (Kubátová, 2012, p. 362). On the subject of reputation in online communities, Wasko and Faraj (2005) argue that an individual's reputation in a collective of people is an important motivation that influences participation in online communities: "Building reputation is a strong motivator for active participation" (Wasko and Faraj, 2005, p. 39). Additionally, the influence of IT is highlighted: "A variety of information technology (IT) artifacts, such as those supporting reputation management [...], are commonly deployed to support online communities" (Ma and Agarwal, 2007, p. 42). One issue regarding the gamification of innovation and ideation is "how the gamification of crowdsourcing leads to an increase in crowd loyalty and idea quality, and therefore supports the development of sustainable (competitive) advantages" (Roth *et al.*, 2015, p. 304). Modern platforms often provide elements such as innovation points, reputation levels and a "hall of fame" for innovative contributors. Accordingly, "Gamification has recently been receiving increased attention in corporate innovation and business research alike" (Roth *et al.*, 2015, p. 300).

The identified facets and representative actions of **Gamification** as the second constituent of the critical success factor **Incentives** are illustrated in Table 51.

Facets	Representative actions
Reputation and Status	• *Active management of individuals' reputation and status on the platform based on set and understood rules; presentation of leaderboards for engaged participants*
Gamification	• *Increased motivation and participation through gamification elements (e.g., idea stocks, innovation points) and auto-graduation mechanisms*

Table 51: *Gamification* as a constituent of *Incentives*[89]

In this study, the innovation experts mentioned the effect of the *Reputation and Status* of individuals on the platform; however, it is important that the rules are understood. The following quotes illustrate this notion:

---

[89] Author's own table.

"We looked very carefully at the stakeholders, what they really cared about, and completely refocused the program, particularly the rewards and recognition aspects. **So we built an innovative points program that we felt were going to drive the right kind of behaviors but also a recognition-based program, a status.***"*

- Expert, Case A

***

*"**Communicate the basis for statuses.** This is an important element. If you've got the reputation in this system,* **how many people actually understand how the reputation works** *within the system?"*

- Expert, Case A

Additionally, the use of *Gamification* elements is highlighted. Some excerpts can illustrate and further enhance the meaning of engaging potential volunteers to participate in firm internal Innovation Contests:

"You already know these things; this is our system, right? **Innovation points, reputation, leader board.** *You can give innovation points, obviously you know about points. Statuses highlight your leaders."*

- Expert, Case A

***

*"**Employees only consent to gamification when the rules are understood.** If it's some sort of black box, you don't know what's going on, that's scary. I don't want to play in that.* **The process must be perceived as fair,** *right?"*

- Expert, Case A

***

*"**Recognize that gamification only enhances.** This is not the silver bullet; you don't put gamification in and say voila. It only enhances the other things you're trying to achieve."*

- Expert, Case A

Subsequently, the critical success factor of **Staffing and Community Building** is shown to represent an additional area of action to create an engaging work environment for Innovation Contests.

4.5.3.5 Staffing and Community Building

The critical success factor of **Staffing and Community Building** consolidates all of the **Staffing** activities of key personnel and of the activities of **Community Building,** which stands for generating an audience and engaging in expert integration.

First, the **Staffing** constituent emphasizes a suitable and best-possible appointment of key personnel and members of several teams and roles for all activities according to the firm internal Innovation Contests, e.g., the implementation team, innovation or campaign managers, evaluators, and experts.

Regarding the state of the art in the literature, "a successful innovation process requires the involvement of highly qualified people" (Schmelter *et al.*, 2010, p. 165). To develop employees in organizations and innovation-relevant roles, McGourty *et al.* (1996) suggest that organizations should be "committed to helping their people develop as innovators. [...] Their development efforts include training programs in teamwork, interpersonal skills, idea generation techniques and managing the innovation process" (McGourty *et al.*, 1996, p. 363).

A complete overview of the facets and representative actions regarding the constituent **Staffing** of the critical success factor **Staffing and Community Building** is presented in Table 52.

Facets	Representative actions
Staffing of Key Personnel	• *Staffing of key personnel (implementation and operations team, specialists/experts, jury members) and persons responsible for the development and enhancement of ideas, along with the facilitation of discussions and feedback; providing training for various roles and groups (moderators, evaluators, jury, campaign manager)*
Innovation Advocacies	• *Creation of a network of volunteers functioning as additional contact persons for all employees, e.g., to respond to unanswered questions and to enhance contestants' participation*

Table 52: *Staffing as constituent of Staffing and Community Building*[90]

---

[90] Author's own table.

In this study, the following excerpt illustrates the importance of the facet of *Staffing of Key Personnel:*

> *"It's not only about employees, because they submit [...]. Overall, those are not really actionable ideas that we can directly implement. We have to take them further and for that, you also need other stakeholders. So, of course, we have the campaign team, campaign sponsor, campaign evaluators, and campaign manager."*

- Expert, Case L

In addition, the establishment of an internal, relatively small network of *Innovation Advocacies* to support the innovation initiative in its local environment is revealed as an important element, as highlighted in literature: "In the office we studied, nearly every person was named as a helper by at least one other person" (Amabile *et al.*, 2014, p. 7). Additionally, similar to inner sourcing in open source projects, contributors "can become coordinators, or 'trusted lieutenants', helping the 'benevolent dictator' manage and coordinate the project" (Stol and Fitzgerald, 2015, p. 65). The innovation experts also highlighted the facet of *Innovation Advocacies* for Innovation Contests as follows:

> *"Every client that we have that has some form of advocacy program or innovation champion program where we have unpaid professionals doing day jobs, but they were also allied to the innovation team. That always boost participation and quality, it makes everything better."*

- Expert, Case A

***

> *"You need touch points in the real world and therefore, we have rolled out and organized what we call an 'Innovation Catalyst Network' that includes people who are enthusiastic and motivated, kind of missionaries, evangelists of innovation, going out in their specific local environment to bring the people together around innovation subjects; to be a first aid for innovation."*

- Expert, Case I

***

> *"OK, so now what we have to do is to keep the mission running. For this, we have now realized other activities [...] which the people get if they provide an idea; direct contacts and an innovation ambassador concept."*

- Expert, Case O

Second, the constituent of **Community Building** consolidates all activities for generating the audience, *inter alia*, to build a network and community of innovators or to integrate experts. This includes the support to allocate, engage, and motivate resources from various departments, divisions and company-wide and to consider various groups such as top performers, blockers, and undecided employees. This constituent also encompasses the importance of collaboration between employees from different areas ("breaking the silos") to develop novel ideas. It also includes the needs and motives of the crowd, e.g., influencing decision making, helping each other, and gaining recognition or satisfaction should be taken into consideration during staffing and selection activities.

In the literature, for instance, Digmayer and Jakobs (2013) describe the essential existence of the online community for Innovation Contests and distinguish different forms of elaboration of contributions and collaboration. According to them, the spectrum ranges from mob-based contests (low collaboration) to community-based contests (medium collaboration) to expert-based contests (high collaboration). Additionally, the simultaneous existence of competitive and cooperative behavior is mentioned in the pertinent literature (Hutter *et al.*, 2011; Bullinger *et al.*, 2010).

The facets and representative actions of **Community Building** as a constituent of the critical success factor **Staffing and Community Building** are given in Table 53.

Facets	Representative actions
Community Development and Collaboration	• *Encourage the creation and development of a community of volunteers for planned and current Innovation Contests and enhance the sharing of knowledge and collaboration within the company*
Composition and Allocation of Resources	• *Ensure the composition and allocation of resources on a corporate-wide level for full utilization of the initiative; support expert search through maintenance of personal profiles and data*
Skills, Expertise and Knowledge Diversity	• *Administration, combining, and exploitation of skills, expertise, and knowledge diversity from different areas of the firm and existing products, services, processes, and businesses*

Table 53: *Community Building* as a constituent of *Staffing and Community Building*[91]

The following quotes illustrate the necessity of the *Community Development and Collaboration* element:

---

[91] Author's own table.

"It's about connecting the people, **creating a network within the company."**

- Expert, Case J

***

"It's the combination of people, the combination of people and digital information and infrastructure. [...] But what's important there is that **people are at the heart of that;** *[...]* **the collaboration with people component and making use of the platform."**

- Expert, Case R

***

" **The final part that we looked at was collaboration.** *Because essentially, it was closed [...], people weren't able to build and improve upon the ideas of others. We felt* **there was a big opportunity there to improve the quality of the content."**

- Expert, Case A

***

"A living community can be your employees [...]. We ask them to gather their insights. We ask them to aggregate the complexity in the market. We ask them to share."

- Expert, Case P

In addition, *Composition and Allocation of Resources* can be presented as second facet that is highlighted as an important element to facilitate **Community Building**. The innovation experts highlighted this point, for instance, in the following quotation:

"What you will experience is that online groups don't necessarily behave as you expect. **The behaviors** *that we see in perhaps more social environments online such as Facebook or Twitter* **do not necessarily correlate to what we see inside our enterprise.** *There are some big differences."*

- Expert, Case A

***

"At work, I have a day job. And any online collaborative activity usually has a business point to it. So people will be more conservative, not necessarily stoking to share. The dynamic is different. We need to understand the social science behind how people collaborate online in a business context so that we can help guide our organizations through this work, **and engineer people to collaborate well and effectively focus on what we really care about."**

- Expert, Case A

***

"There is synchronization of the defined user groups. **We already had a user address book, a global address book**, *and we don't want to make it a second one."*

- Expert, Case M

In addition, the exploitation of *Skills, Expertise and Knowledge Diversity* from various areas of the firm, the use of a cross-functional and cross-regional information flow, and the use of knowledge diversity and expertise involving existing products and services determines the success of initiatives. The following quotes illustrate these needs:

*"***The expertise and [...] the skills***, of course; we often face completely new business fields. We are not experts in them. Should we go this way and go into, let's say, a red ocean, maybe? This is a challenge."*

- Expert, Case K

Below, **Monitoring** as an additional area of action for the creation of an engaging work environment for Innovation Contests is presented.

4.5.3.6 Monitoring

The critical success factor of **Monitoring** focuses on the effective monitoring and presentation of several aspects, including **Progress Monitoring** (focusing on, *inter alia*, progress) and **Benefits Monitoring** (focusing on, *inter alia*, the financial side). Accordingly, this distinction is similar to the differentiation in the idea community scorecard by Blohm *et al.* (2011b), who formulate indicators based on four dimensions: The financial dimension, the process dimension, the customer dimension, and the perspective of external actors. However, the latter two—the involvement of customers and/or external actors—are not relevant to this study's concentration on internal contestants.

First, the constituent of **Progress Monitoring** consolidates all monitoring activities for process execution, progress monitoring, and monitoring of information objects (campaigns, contributions, participants, etc.). Therefore, input and output measures or performance indicators, for instance, stage gates and idea quantity indicators, must be considered.

An overview of **Progress Monitoring** as a constituent of the critical success factor **Monitoring** is depicted in Table 54. For all facets, it is important to provide the monitoring results to interested employees so that potential participants and contestants are made aware of achieved progress.

Facets	Representative actions
Monitoring of Process Execution and Progress	• *Monitoring of the process execution and achieved progress supported by a priori defined key performance indicators (KPIs)*
Monitoring of Information Objects	• *Monitoring of central information objects (#ideas, #participants, #voting's, campaigns) and stage-gates*
Customizable Reporting	• *Ensuring the availability of the monitoring results (customizable) for various stakeholder and expert groups*

Table 54: *Progress Monitoring* as a constituent of *Monitoring*[92]

The following quotes illustrate the importance of the **Progress Monitoring** element. First, *Monitoring of Process Execution and Progress* must be highlighted:

> *"Our management is asking us about KPIs. We have approved budget. We have a number of ideas. One of the KPIs is actually,* **'how long does it take to grow an idea'***, an* **initial idea***; this first thinking, this first thought of one person to an idea, which is presentable at the management board and which is understandable?"*

\- Expert, Case P

Additionally, *Monitoring of Information Objects* is revealed in this study as an important element, as the following quotes illustrate:

> *"****Interesting was to measure our efficiency and set new goals*** and also very important for the people that are managing innovation programs."*

\- Expert, Case J

***

---

[92] Author's own table.

"All of these things are actual useful barometers of the level of activity, the level of take up, the adoption, the interest."

- Expert, Case R

In addition, the availability of *Customizable Reporting* as an important facet can be found in various text passages, as seen below:

"It's very important to update the employees[93] with progress and updated reports about how the previous campaigns went."

- Expert, Case L

In addition to **Progress Monitoring**, the second constituent of the critical success factor **Monitoring** is titled **Benefits Monitoring**. Here, both benefit considerations (*inter alia,* the quality of ideas and contributions) and generated business values should be assessed and monitored. *Benefits Monitoring* considers the benefits of the innovation initiative that result not only from monitoring generated values but also from innovation capacity and efficiency.

The identified facets and representative actions of **Benefits Monitoring** as a constituent of the critical success factor **Monitoring** are presented in Table 55. Again, the public availability of the monitoring results that consider the benefits is important so that potential contestants' perception of the achieved outcomes increases.

Facets	Representative actions
Benefits Monitoring	• *Monitoring of the benefits (and costs and investments) of the initiative through the consideration and identification of generated values (in the pipeline), the value propositions associated with ideas and concepts, and measuring innovation capacity and efficiency*
Health Checks, Reviews and Recommendations	• *Health checks and assessments, including recommendations for optimizing the entire initiative and Innovation Contests activities and reviewing the qualitative and quantitative results after defined periods of time*

Table 55: *Benefits Monitoring* as a constituent of *Monitoring*[94]

In this study, *benefits monitoring* can be recognized in the following quotes:

---

[93] For the sake of clarity, the author replaced "them" with "the employees".
[94] Author's own table.

"Go through the participation and look at the quality, look at how your organization is using the content that you're driving, and help you refocus the program in order to better meet your goals and expectations. [...] We feel it has benefit. We feel it has value."

- Expert, Case A

***

"We run activities for 5 years. What are the results? What can be measured? Where is the value? Where is the EBIT?"

- Expert, Case K

The latter element, *Health Checks, Reviews and Recommendations*, represents the execution of health checks, reviews and assessments of the innovation initiative, including recommendations for optimization and the qualitative and quantitative review of the results achieved. These elements can be identified in various text passages, including the following:

"This is a much more significant set of changes that were required. We did a health check [...]. We looked at the program in its entirety, and what we found [...] were a wide variety of changes we could make which would improve things."

- Expert, Case A

Along with the areas of actions on the strategic and initiative levels, as previous introduced, additional critical success factors of an engaging work environment are located on the contest level as described below.

4.5.4 Decomposition of support factors on the contest level

On the contest level, areas of action for the creation of an engaging work environment that builds organizational support are identified before, during and after the execution of Innovation Contests and are embedded in the innovation initiative. The matching critical success factors are **Campaign Alignment**, **Preparation**, **Execution**, and **Transition**. A detailed presentation of these critical success factors, their constituents, and facets is provided in the texts set forth below.

4.5.4.1 Campaign Alignment

The critical success factor of **Campaign Alignment** refers to a coordinated and agreed alignment between the setup of the Innovation Contest and its environment. Increased consideration of **Campaign Alignment** ensures that running Innovation Contests fulfill expectations (e.g., the needs of the managers or the business) and solve existing business challenges. Additionally, the connection between the Innovation Contest and the business's problem, needs and wishes, campaign focus, and fit between the challenge and potential contestants' competencies should be considered. Following a suggestion in the literature, "individuals' idea generation efforts are clearly directed toward specific areas and problems. These companies also focused on integrating technology and business objectives, thereby helping to ensure that researchers maintain an applied orientation" (McGourty *et al.*, 1996, p. 363). The **Campaign Alignment** (i.e., the alignment between the campaign and its environment) must be considered separately compared to the constituent **Organizational Alignment** described above, which focuses on the alignment between the entire initiative and the organization (and its strategy, goals, etc.).

With respect to the pertinent literature on, *inter alia*, the audience selection for Innovation Contests, "the organizer also indicates the interesting target group of participants. Literature identifies a distinction between unspecified target group, i.e. participation is open to everybody and a specific target group, when participation is e.g. limited to a country or qualified by age or interest" (Adamczyk *et al.*, 2010, p. 3). Also Schulze *et al.* (2012b) focus on the importance of having the demand-abilities fit represent (preferably) a close match between the task requirements and the competencies of individuals; this could arise as an important challenge for organizations (Hetmank, 2013).

The facets and representative actions related to the critical success factor of **Campaign Alignment** to the organizational environment are provided in Table 56.

Facets	Representative actions
Alignment of Innovation Contests	• *Establish targeted Innovation Contests to ensure the fit of participants' skills and knowledge to the challenge ("demand-fit"), thus avoiding irrelevant ideas and inappropriate behavior; consideration of defining criteria tailored to the selected audience and crowd size*
Tailoring of Innovation Contests	• *Tailoring Innovation Contests according to the pertinence, importance, and priority of the challenge and strengthening the focus and interest of campaigns (e.g., hot topics, new sponsors and experts)*

Table 56: Representative actions for the critical success factor *Campaign Alignment*[95]

---

[95] Author's own table.

The *Alignment of Innovation Contests* is revealed as the first important facet during the data analysis, as illustrated in the following excerpts:

*"**What we really do is strive to do the campaigns for the topics that matter to the business**. A, to better serve the business; B, to increase the chance that ideas really get implemented because we ideate on topics that they already have on the target; but maybe don't know yet how to fix it."*

- Expert, Case L

***

*"Maybe this isn't great that all of the innovations topics were very similar. We're engaging the same group of people every single time. So maybe **we needed some more diversity to see if we could bring in different parts of the organization**."*

- Expert, Case A

***

*"Again, the **main attention of the whole program is to focus on campaigns with a very specific question that is always coming from the business** and is time limited. But most important, we facilitate this to the rest of the company."*

- Expert, Case L

Accordingly, the facet *Tailoring of Innovation Contests* states the importance of adjusting the setup of upcoming Innovation Contests to the pertinence, importance and priority of the challenge:

*"**I need different kinds of campaigns**. Maybe, I have top-level campaigns that come straight from the CEO, but I would also like to support peer campaigns. Campaigns that are smaller, or just a department leader [...] reaches out to people across the company. **That's a pretty different kind of campaign**."*

- Expert, Case A

Subsequently, **Preparation** as an additional area of action is highlighted.

4.5.4.2 Preparation

As the next critical success factor, the **Preparation** of Innovation Contests incorporates the achievement of a common understanding; *inter alia*, by either considering expert involvement, and scheduling or determining the degree of innovation for every challenge. The **Preparation** of the Innovation Contest is assessed as the determining factor for the execution and the achieved and submitted contributions. Potential contributors want to understand what the initiative expects from them and what they need to do (or what they can do) during the various activities. Adamczyk *et al.* (2010, p. 3) illustrate that "usually, the organizer dedicates the contest to a specific topic; details of which vary extensively. The topic indicates specifically of the task/topic [...] and the desired degree of elaboration" (Adamczyk *et al.*, 2010, p. 3).

In the pertinent literature, especially on online communities, "the establishment of mutual understanding to comprehend conversations and knowledge distribution is inevitably more difficult than face-to-face communication in a small group" (Ma and Agarwal, 2007, p. 43). The formulation of a precise problem statement is an important determinant for the success of Innovation Contests; this issue primarily influences participation (Yang *et al.*, 2009). "The seeker has to be able to provide a clear description of the problem. If the problem is highly complex with ill-defined interfaces, it is not suitable for an Innovation Contest" (Terwiesch and Xu, 2008, p. 1530). Afuah and Tucci (2012) demand that the task should be clearly delimitable and characterized by a low level of implicit knowledge and complexity.

An overview of the identified facets and representative actions for the critical success factor of **Preparation** is given in Table 57.

Facets	Representative actions
Problem Framing	• *Innovation Contest creation and preparation with the involvement of experts; achievement of a clear, common understanding of, e.g., the mission, goal-setting, problem statement, and success criteria for every contest*
Definition of the General Conditions	• *Definition of all general conditions for the contest execution, including a staged campaign schedule with a fixed period of time or determination of the degree of innovation*

Table 57: Representative actions of the critical success factor of *Preparation*[96]

Excerpts for the facet of *Problem Framing* are found in various passages, including the following:

---

[96] Author's own table.

> *"I think everyone agrees on a few good practices that you need to follow to be successful. I also think that* **the content of your campaign is really critical** *and actually, that's something that we are struggling with. And I am wondering what kind of support you can give* **to translate strategic, very abstract, strategic goals to very specific questions in a campaign.***"*

\- Expert, Case A

***

> *"The real starting point was then if* **we switched to a dedicated focus, what we call competitions, idea contests where we have a clear problem statement and success criteria** *that we have agreed with sponsors."*

\- Expert, Case I

***

> *"Before you start the campaign even, as I think Einstein said that '95 % of taking in one hour left of his life to spend into problem definition and only five percent of the time on the solution.' So this is true also for campaigns, launching campaigns.* **You should spend a lot of effort to do this in terms of problem framing. What is the real problem?***"*

\- Expert, Case I

Additionally, the facet of *Definition of the General Conditions* involves defining all the important general conditions before beginning Innovation Contests. The Innovation experts stated the importance of this facet below:

> *"***Look at that planning and alignment aspect.*** *So what is it you care about, where are you trying to head over what kind of time frames? Well, then we'd also roll our sleeves up and actually work with you on the questions, focusing on those in an engaging way. So look at the questions that have worked well and look at the questions that don't work so well."*

\- Expert, Case A

Below, details about **Execution** as an additional area of action that is needed to obtain an engaging work environment are presented.

4.5.4.3 Execution

The critical success factor of **Execution** consolidates support during the **Ideation, Discussion**, and **Evaluation** stages of running corporate Innovation Contests.

First, the constituent of **Ideation** is informed by the organization's supportive ground-work for generating, enriching and detailing ideas, *inter alia*, through the organization and moderation of Innovation Contests, the administration and documentation of idea drafts and the building of an idea pool.

The identified facets and representative actions of **Ideation** as a constituent of the critical success factor of **Execution** are presented in Table 58.

Facets	Representative actions
Idea Generation	• *Support the effective creation and administration of idea drafts (e.g., categorization, prescreening); ensure the building and administration of the idea pool (idea description, ideators and meta information, tags, categories, attachments)*
Information and Idea Flow	• *Encouragement of a cross-functional (and cross-regional) idea exchange, expertise sharing, and information flow*
Organization and Moderation	• *Organization of all aspects for offering of a supportive groundwork for idea enrichment and detailing, especially through organizing and moderating Innovation Contests*

Table 58: *Ideation* as a constituent of *Execution*[97]

The facet of *Idea Generation* to manage and organize all upcoming ideas and additional contributions is important, as set forth in the following excerpts:

> *"Accepting they are in a digital environment,* **people want to be able to generate and contribute at the time that best suits them.** *We have our employees who are consumers, as all of our end-customers are. And we need to think in innovation terms, we need to think in the same way. About* **how do we give them the tools to contribute and to participate that best suits them?** *Where are we going to get the best contributions?"*

- Expert, Case R

***

---

[97] Author's own table.

"The second aspect is about encouraging our different viewers to engage into innovation activities. When we go back to the discipline of innovation [...] **The first step is being able to generate and capture ideas.***"*

- Expert, Case J

***

"In the end, we have the ideation process **where we find out ideas on the product or on the service level** *for these defined search fields. So, this is more or less the fuzzy front end and [...] a very common idea management system."*

- Expert, Case N

In addition, an active *Information and Idea Flow* is highlighted as important:

"What was interesting for us is the value proposition of innovation management software mainly because **they are allowed to process those ideas and opinion within the group of employees.** *That pulled all of us to manage the workflow from ideation to implementation."*

- Expert, Case J

Additionally, the analysis reveals the element of *Organization and Moderation* as important. Moderation activities are especially important to offer support to all contributors during the ideation step. Moreover, moderation can help operate the community and its behavior through a rapid, precise intervention. Chen *et al.* (2011) have shown the aspect of moderation in online communities and its influence on the quality of generated content.

Second, the constituent of **Discussion** focuses on questioning the published ideas and further developing ideas by not only collecting comments and improvement suggestions but also maintaining initial ratings and crowd evaluations. Table 59 illustrates the facets and representative actions of **Discussion** as a constituent of the critical success factor **Execution**.

Facets	Representative actions
Continuous Idea Development	• *Ensure idea development by scrutinizing, advertising, and further developing ideas, e.g., by finding contributors or obtaining comments and improvement suggestions*
Crowd Evaluations	• *Engage initial ratings of published ideas through crowd discussions, contributions, and voting*

Table 59: *Discussion* as a constituent of *Execution*[98]

First, the element of *Continuous Idea Development* helps improve the quality and elaboration of ideas. The following excerpts illustrate its importance:

*"With the tool, we had **the possibility that everybody in the organization is able to contribute to an idea**. That's a first very important step, which was a very great success we had. In the beginning, a lot of ideas came up."*

- Expert, Case P

***

*"An idea is actually a complicated, complex construct. It's not just an idea and some words on it. **Everybody comes from a different angle. Everybody has a different perception on the idea.**"*

- Expert, Case P

***

*"They can give proper questions, **give proper feedback to help the idea generators lift up their idea**, because I'm not sure if that's your same experience, but what I see is that the idea generators often have a far bigger idea than they write down. They are a bit lazy, so they do it quick. They write down the idea, but if you question them and say hey, have you thought about this? What's your solution for this? **Then they will write it down and the idea improves and it becomes a more actionable idea.**"*

- Expert, Case L

In addition, the data analysis reveals opinion formation, expressed as *Crowd Evaluations*, as an important element.

Third, the constituent of **Evaluation** includes an engaging environment for the entire evaluation process, *inter alia*, supporting voting by various stakeholders (e.g., individuals or groups of experts) and assessing and selecting ideas based on defined criteria.

---

[98] Author's own table.

Adamczyk *et al.* (2010, p. 4) mention that "once submissions are made, their evaluation can be made along to three basic pathways which can be freely combined: Self-assessment by the participant, peer review by the (other) participants of the Innovation Contest and evaluation by a jury of experts" (Adamczyk *et al.*, 2010, p. 4). Furthermore, idea selection is assessed as a challenging and important step in the ideation phase (Yuecesan, 2013). Riedl *et al.* (2010) argue that evaluation mechanisms with more subtle distinctions are more precise than simple mechanisms.

The identified facets and representative actions of **Evaluation** as a constituent of the critical success factor of **Execution** are provided in Table 60.

Facets	Representative actions
Idea Evaluation	• *Support the evaluation of ideas based on defined quality*
Idea Selection	• *Ensure a transparent and comprehensible idea selection process, including feedback to the idea provider based on understandable evaluation criteria*

Table 60: *Evaluation* as a constituent of *Execution*[99]

First, this study confirms the importance of the *Idea Evaluation* step. The following quote from an innovation expert illustrates the need for this step:

"Think about who the evaluators are because some people just only see the downside of a new idea and that is not always in the interest of the whole company. **So use different evaluators, have them to sit them together in a room, evaluate together, create and facilitate a discussion and then make a decision."**

- Expert, Case L

In addition, the *Idea Selection* element is seen as an important facet of support:

"Be prepared to kill a lot of ideas. *There will be a lot of ideas generated. We saw a lot of ideas coming in and not all ideas are that groundbreaking, so please be prepared to kill the not-so-good ideas.* **Take into account, that you have a good explanation so that the idea generator accepts the idea's deactivation."**

- Expert, Case L

Transition is the last identified area of action to create an engaging work environment that supports the success of firm internal Innovation Contests.

---

[99] Author's own table.

4.5.4.4 Transition

On the contest level, the success factor of **Transition** consolidates activities for managing the post-contest process and is relevance following an Innovation Contest. It incorporates issues related to the transition of selected ideas to the various steps of the post-contest process (e.g., concept creation, prototyping, idea implementation).

An overview of the facets and representative actions related to the critical success factor of **Transition** is presented in Table 61.

Facets	Representative actions
Idea Transition	• *Engagement of idea brokerage, quick transfer of feasible ideas into business units or new business development, effective routing, continuous flow of ideas, transforming ideas to realization*
Project Initiation	• *Project initiation and concept management, clear transition into concepts and projects, obtaining early 'buy-in' from business to implement ideas*
Business Case Development	• *Support for the development of business cases and business plans, solution modeling and prototyping*

Table 61: Representative actions of the critical success factor of *Transition*[100]

To be more concrete, *Idea Transition*, *Project Initiation*, and *Business Case Development* are the relevant representative actions. With respect to transitioning the results to the subsequent steps of the process in a cooperative context, several experts must be involved, with an efficient *Idea Transition* as the first step:

> "*You have people that know the market and are able to create a vision. You have here in the company different* **people that know to operate our work facilities and how best to apply innovation.** *You have here people who know best which business model should be applied to an invention to make the maximum impact. You have* **people who know how to develop new products or new solutions** *or to look for the solution that are outside the company.*"

- Expert, Case J

***

---

[100] Author's own table.

"We have proven that what is coming in is acted on; going back to the people coming up with the idea, and that it's implemented. So it's a kind of closure between ideas and execution, which has proven extremely important for people."

- Expert, Case Q

Second, reliable *Project Initiation* was revealed by the data analysis as an important facet and as part of the critical success factor of Transition.

"We thought what to do next, with an idea as a project defined. We want to go further to the back end [...]. We have the idea creation process. We have the process for the research to make building blocks. And then we have the process for developing new products, developing new equipment, and maintaining products."

- Expert, Case M

Finally, a structured *Business Case Development* as a facet of the Transition success factor was highlighted, as illustrated by the following quotation:

"After approval, funding is ready and colleagues from solution development care for development, positioning, and maintenance."

- Expert, Case P

Accordingly, the presentation of the areas of action as aspects of the creation of an engaging work environment for participation in firm internal Innovation Contests is finally conducted. A summary of Study B is presented below.

4.6 Summary of Study B

The second study *("STUDY B: Exploring an engaging work environment for Innovation Contests")* and its results reveal twelve areas of action and 21 subjacent constituents of an engaging work environment for the organizational support of Innovation Contests. These areas of action are "Strategic Sponsorship", "Vision and Strategy", "Visibility", "Process and Transparency", "Communication", "Incentives", "Staffing and Community Building", "Monitoring", "Campaign Alignment", "Preparation", "Execution", and "Transition". The recipients of all the activities within the identified areas of action are the group of employees as potential participants in innovation activities.

In chapter 5, an overall summary and the resulting practical and theoretical implications are provided as the final part of this dissertation.

5 Summary, conclusion and outlook

This chapter provides a summary in section 5.1, followed by the conclusion in section 5.2. Finally, an outlook is given in section 5.3.

5.1 Summary

Over the last decade, firm internal Innovation Contests as a popular and frequently realized application for facilitating innovation activities have gained attention both in theory and in practice (Adamczyk *et al.*, 2012; Bullinger and Moeslein, 2010). Embedded in this context, this doctoral dissertation focuses on the firm's activities and practices that support the optimal utilization of Innovation Contests within a corporate environment. The overall aim was an investigation of the impact and institutional design of a supportive and facilitating work environment, along with organizational support for employees' *motivation, affective organizational commitment*, and intention to participate in Innovation Contests because active participation is important for the overall performance and success of such activities (Zheng *et al.*, 2011; Leimeister *et al.*, 2009). To investigate this phenomenon in a real-world context, two studies were executed. In the first, an explanatory study, an empirical investigation and statistical analysis of the influence of work environment perceptions and organizational support on employees' attitudes and desired behavior was conducted. That study used an online questionnaire and the structural equation modeling approach. In the second, exploratory study, an in-depth analysis and identification of the aspects of an engaging work environment and the constituents of organizational support, tailored to the context of firm internal Innovation Contests, was provided. According to the Componential Theory of Creativity and Innovation in Organizations, the aspects that were identified might commonly shape employees' positive work environment perceptions. This research was planned and executed in cooperation with a leading firm specializing in innovation management software and corporate Innovation Contests. Using their products, consulting and customizing services, approximately 200 customer-firms across Europe and worldwide have successfully implemented corporate Innovation Contests. The findings of this dissertation provide several theoretical contributions, help managers support the utilization of Innovation Contests in the best way possible, and make suggestions for further research.

As stated in the introduction to this doctoral dissertation, this thesis addresses Innovation Contests that increase internal labor for the development of ideas and innovations; it also addresses the work environment as an important driver of employees' behavior related to their participation in such activities. The corresponding research questions for this dissertation and its two interrelated studies are as follows: "What is the influence of different work environment perceptions on employees' affective organizational commitment and on their motivation and intention to participate in firm internal Innovation Contests?" (Research question 1); and "Which areas of action (critical success factors, constituents and facets) exist that represent aspects of an engaging work environment in firm internal Innovation Contests?" (Research question 2). These questions represent major challenges for firms because employees' active participation and high motivation are assessed as critical factors for the overall performance of such activities (cf. Frey *et al.*, 2011; Zheng *et al.*, 2011). Moreover, Franke *et al.* (2014, p. 1) highlight that "the success of the tournament rests primarily on the number of participants attracted". Accordingly, this dissertation addresses a practical, highly relevant phenomenon because firms often struggle with the implementation and utilization of firm internal Innovation Contests in which a broad range of aspects must be considered.

The verification of causalities and the achieved results, based on an investigation and survey research within one large, multi-divisional company in Germany, offer interesting insights and an extension of the existing body of knowledge. The aim of this quantitative investigation was to analyze the impact of diverse work environment perceptions, illustrating the social environment within the firm and on a related note, the application of the Componential Theory of Creativity and Innovation in Organizations to the context of firm internal Innovation Contests. The empirical findings offer a better understanding of the hypothesized relationships among several work environment perceptions (both positive (*organizational encouragement, supervisory encouragement*) and negative (*organizational impediments, and workload pressure*)) and employees' *affective organizational commitment, motivation,* and *participation intention* related to firm internal Innovation Contests.

The survey instrument was web-based, whereas the corresponding scales were drawn from the existing literature and adapted to the study context. Additionally, diverse control and demographical variables were collected. The invitation to participate in this study was sent to approximately 750 employees of a German DAX 30 telecommunications company subunit responsible for group-wide development of the company's product and service portfolio. One euro was donated for each completed survey. One hundred and fifty-four responses were usable for the data analysis. The descriptive statistics indicate a broad range of respondents from the surveyed unit. The quality of the questionnaire structure (mean: 2.15 on a five-point Likert scale from 1 to 5) and the clarity of the questions (mean: 1.86) were assessed as good.

The confirmatory factor analysis indicates suitable factor reliabilities for all scales: *Organizational encouragement* (0.862), *supervisory encouragement* (0.935), *organizational impediments* (0.832), *workload pressure* (0.768), *affective organizational commitment* (0.913), *motivation* (0.879), and *participation intention* (0.947). Different tests intended to reveal any common method bias indicate that the collected data are not biased. Summarizing the main findings, SEM with a suitable goodness-of-fit reveals several significant effects between the investigated variables. More specifically, *motivation* is affected by *organizational encouragement*. *Organizational encouragement* and *supervisory encouragement* influence *affective organizational commitment*. Furthermore, it is shown that employees' *participation intention* is affected by their individual *motivation* and their *affective organizational commitment* to the organization. With the provided results, a generalization of the findings based on "incorporated causal paths and the identification of the collective strength of multiple variables" (Creswell, 2013, pp. 13–14) is possible because the quantitative method of this survey seeks "to discover relationships that are common across organizations and [...] to provide generalizable statements about the object of study".

Compared to the positive relationships, there are relatively few studies on either *organizational impediments* or *workload pressure* (cf. Amabile *et al.*, 1996). This study's attempt to validate significant applicable effects was not completely successful. Additional interpretation might be necessary for the results on *workload pressure*. The results reveal its small-but-positive instead of (as assumed) negative relationships with *affective organizational commitment* and *participation intention*. The results of this thesis in this area can be explained to some extent by Amabile *et al.* (1996, p. 1161), who highlight the following observation: "The evidence that does exist suggests seemingly paradoxical influences. Some research has found that, although workload pressures that were considered extreme could undermine creativity, some degree of pressure could have a positive influence if it was perceived as arising from the urgent, intellectually challenging nature of the problem". Or, as additionally explained: "One explanation for inconsistent findings on stress-performance relationships is that there is 'good' stress as well as 'bad' stress" (LePine *et al.*, 2005, p. 764).

Moreover, the moderation analysis reveals that when employees' *affective organizational commitment* is already at a high level, the organization's encouragement of their creativity and innovations (organizational encouragement) significantly increases their *participation intention* related to firm internal Innovation Contests. For the relationship between *motivation* and *participation intention*, there was no moderation effect, indicating that Innovation Contest organizers should be aware of *organizational encouragement* as a prerequisite of employees' *motivation*, which in turn is a predictor of *participation intention*. Additionally, the mediation analysis reveals that although there is no direct effect from *organizational encouragement* to *participation intention*, there is an indirect effect on this relationship through *motivation*.

The outcomes of the qualitative part offer interesting additional insights. In line with the motivation and introduction of this thesis, the aim of study B was to identify the areas of actions representing aspects of an engaging work environment that might affect employees' positive work environment perceptions and ultimately (as proven in Study A), increase their work-related behavior related to participation in firm internal Innovation Contests. To investigate this phenomenon in a real-world context, an exploratory study design based on a qualitative content analysis and inductive reasoning was chosen. This approach included a systematic model for qualitative research that contained seven steps. To ensure the quality of this research, adherence to diverse quality criteria (objectivity, reliability, internal validity, external validity, and utilization) was ensured.

To identify the relevant aspects of an engaging work environment, a cross-case comparison that considered the best-practice presentations given at Germany's annual Innovation Managers Forum was conducted. Overall, presentations from 18 European different firms with between 3,700 and more than 400,000 employees were considered. All the companies rely on an identical, but customized software solution for conducting firm internal Innovation Contests and have already reached a mature state of implementation. All the best-practice presentations were recorded on video and prepared for the purpose of this study, leading to more than 500 minutes of video footages and more than 100 pages of transcripts.

The data analysis was conducted based on defined coding rules and in four separate analytical steps. Here, the coding scheme for structuring the data was developed in an inductive way and refined in diverse revisions and pilot testing. Overall, 961 passphrases were found during the initial coding, grouped into 21 constituents and 12 critical success factors representing aspects of an engaging work environment.

The main results lie in the study's presentation of the totality of the areas of action in the implementation state, during which corporate Innovation Contests are fully developed and integrated into the organization. On the highest level, the framework offers an overall presentation of company-wide organizational support for firm internal Innovation Contests. On the level of the identified critical success factors, this study clusters the aspects of an engaging work environment that influence employees' perceptions into 12 areas of action, namely, "Strategic Sponsorship", "Vision and Strategy", "Visibility", "Process and Transparency", "Communication", "Incentives", "Staffing and Community Building", "Monitoring", "Campaign Alignment", "Preparation", "Execution" and "Transition". A clarification of these aspects is provided by representative citations from the innovation experts' daily work on the success of firm internal Innovation Contests. At this point, a more complete and nuanced view of organizational encouragement of participation in firm internal operated Innovation Contests has emerged. This study's originality arises out of the fact that this is the first work to provide an integrative perspective and complete representation of the critical success factors in this field.

The conclusion and presentation of the practical and theoretical implications are provided in the next section.

5.2 Conclusion

In conclusion, the investigation of the causal relationships between the analyzed latent variables supports existing theory on work environment perceptions as an important driver of employees' innovative behavior (cf. Amabile and Pillemer, 2012; Amabile, 1997), expressed by their intention to participate. Theoretically, the results disclose the adoption of an existing theory (the Componential Theory of Creativity and Innovation in Organizations) as an explanatory model and demonstrate causalities in a novel field of application, namely, the case and context of firm internal Innovation Contests. Additionally, the findings extend the body of knowledge on work behavior, especially in the area of extra-role behavior (cf. Rhoades and Eisenberger, 2002). The work contributes to literature in which "research reveals a strong theoretical and empirical connection among organizational conditions, employee motivation, and performance" (Mudambi *et al.*, 2007, p. 443). As summarized in the previous sections, some relationships are already known and depicted in the literature on a different or more generic level within organizations. Nevertheless, important influences of work environment perceptions on Innovation Contest participation were shown; this is a research topic that previously was neglected. In particular, the influence of *organizational encouragement* as the only perception that affects both employees' *affective organizational commitment* with the organization and employees' *motivation* in the context of Innovation Contests is validated and should be regarded as a determinant of major interest for further research. Furthermore, *affective organizational commitment* and *motivation* significantly influence employees' *participation intention* and are seen as a suitable and reliable predictor of actual participation.

In addition, the results of the qualitative content analysis, which are based on the best-practice presentations of the innovation experts at the annual innovation managers' forum, expand the literature on Innovation Contests by offering an extensive representation of important areas of action ranging throughout the firm's hierarchy. Considering its theoretical implications on a general level, the results offer an enhanced understanding of an engaging work environment and the subjacent aspects of organization-wide support tailored to the area of Innovation Contests within firm boundaries through a more in-depth investigation. This work sheds light on the abstract construct of an engaging work environment in the context of Innovation Contests and offers a description of how the identified areas of actions can be designed. The results of this exploratory study serve as an important source of insight into the engagement of firm internal employees (a previously neglected source of ideas for innovation purposes) as contestants in firm internal Innovation Contests.

Considering the practical implications, first, the certainty and importance of *organizational encouragement* and *supervisory encouragement* as important drivers for desired behaviors in Innovation Contests will help innovation managers and other responsible for such innovation activities. Because the value of *organizational encouragement* and *supervisory encouragement* for innovative tasks in the context of Innovation Contests is verified, managers should promote the various characteristics of *organizational encouragement* (i.e., encouraging mechanisms for developing novel ideas, active idea flows, and fair judgments) to improve individuals' perceptions in this area and shape the image of the company.

Second, employees' *motivation* is shown to be an important determinant of the success of firm internal Innovation Contests in terms of increasing employees' participation, which is subsequently assessed as a reliable predictor of real participation (Bateman *et al.*, 2011). Whereas the impact of *motivation* is verified in several studies on Innovation Contests (cf. Zheng *et al.*, 2011; Leimeister *et al.*, 2009; Frey *et al.*, 2011), this work extends the body of knowledge because it is one of the first studies dedicated to Innovation Contests within a corporate environment. In particular, those responsible for the utilization of Innovation Contests within firm boundaries must be aware of the strength of motivation as an important determinant and antecedent of employees' *participation intention* in modern innovation activities.

Third, *affective organizational commitment* as a form of identification with the organization and its influence on employees' *participation intention* in firm internal Innovation Contests is verified. This issue has seldom (or never) been investigated in this context, although its impact is not very strong compared to employees' *motivation*. Considering the contrast between *motivation* (doing something triggered by personal motives) and *affective organizational commitment* (doing something to help the organization) as detailed in section 3.2.1, the results indicate a stronger influence on *participation intention* by employees' *motivation*, thus reflecting the involvement of self-interest and personal motives. This finding gives practitioners an additional adjustment screw for fine-tuning the firm's support for firm internal Innovation Contests.

With respect to the practical implications of the qualitative part of this work and the identified areas of action to create an engaging work environment, this dissertation offers a set of 21 constituents to build an engaging work environment that might help managers to better understand the relevant aspects of organizational support for firm internal Innovation Contests. To that end, this research focuses on the institutionalization step of firm internal Innovation Contests, in which the organization's approach and activities are both consolidated and institutionalized (Chiaroni *et al.*, 2011). In this step, fully established organizational support is necessary. Through the consideration of experiences and knowledge of a large group of innovation experts working on the utilization of corporate Innovation Contests at a high level of maturity and an advanced expansion stage, coverage of the broad range of relevant aspects creating a work environment that supports Innovation Contests can be ensured. On a higher level, the provided framework can serve as a blueprint for generating strategies and development phases and evaluating the actual state of the implementations of firm internal Innovation Contests. Organizations and consulting services can use the framework in practical settings by including the areas of action described in the framework in their implementation project setups. Based upon this approach, a more change-oriented approach can be applied to achieve the desired states within the aspects of an engaging work environment. In terms of transferability, the use of the framework could be helpful, especially in firms with similar characteristics (group or large firm, a large number of employees and prospective participants, specified roles in innovation management and innovation activities), and perhaps for the adoption or differentiation in terms of the content-related design of single success factors.

On a more concrete level, the findings, as presented in the results of Study B, offer several types of assistance through a precise description of the relevant facets and representative actions combined with offering direct quotations to increase comprehensibility and transparency. This study helps increase the understanding of core areas upon which managers should focus when planning to introduce Innovation Contests in their business units. Accordingly, the detail level of the representation of single areas of action could be a starting point for the development of target situations and goals that firms would like to define for better organizational support and a more engaging work environment for Innovation Contests. The identified areas of action and aspects can be used as a guideline by large firms. Thus, an increase in the project success can be achieved and impediments can be minimized.

5.3 Outlook

Further research on Innovation Contests should investigate several issues. With respect to the limitations, first the focus on employees' participation intention in general in this thesis could be mentioned. Here, related future research might differentiate between behavioral actions, e.g. between reading, posting, and moderating actions (cf. Bateman *et al.*, 2011) and investigate, if the influences of an engaging work environment are similar or different. Second, although behavioral intention is assessed as reliable predictor of real participation (Zhou, 2011; Kim *et al.*, 2008), one limitation lies in the fact that this project employs participation intention as latent variable. Furthermore, this study's concentration on best-practice presentations as a primary source of information should be mentioned, although this data source was suitable for this exploratory study and similar sources have been used in previous management studies. Nevertheless, further evaluation of the findings, e.g., by conducting expert interviews, could be advisable.

Additionally, the magnitude of the critical success factors and constituents of an engaging work environment are not further elaborated because of the qualitative study design. Nevertheless, they are essential for the successful implementation of firm internal Innovation Contests in general and should be investigated in future studies. Therefore, the development of scales representing the areas of action might be one objective to consider.

Furthermore, one promising perspective could be obtained by adopting a resource-based view (cf. Wernerfelt, 1984) or a competence-based view (cf. Dutta *et al.*, 2005; Collis, 1994) with respect to the topic of this thesis. Wernerfelt (1984) calls for analyzing the firm more from the resource side because attractive resources (resources with difficult-to-imitate properties that make them difficult for competitors to copy) will lead, for example, to diversification and therefore to a competitive advantage. "A firm's resources at a given time could be defined as those (tangible and intangible) assets which are tied semipermanently to the firm" (Wernerfelt, 1984, p. 172). The resource-based view (RBV) of the firm postulates that it has unique resources and capabilities that primarily determine a firm's increased competitiveness. "Research on RBV is about the use of assets, skills, abilities and knowledge within the firm. The emphasis is on internal resources available and developed within the firm" (Coates and McDermott, 2002, p. 435). Following Coates and McDermott (2002), these resources must fulfill several conditions. They must be difficult to imitate, asymmetrical spreading over firms, and provide opportunities (cf. Coates and McDermott, 2002, p. 436). Future studies could examine the question of how a specialized configuration of resources and competences must be included in the revealed areas of action. Here, the RBV can provide an informative and suitable theoretical perspective for the identification and structured presentation of the abilities of the firm's internal resources.

Another interesting avenue for further research might be to adopt a microfoundation perspective on the investigated phenomenon of an engaging work environment for firm internal Innovation Contests[101]. Adapting the concept of microfoundations to this study, the identified areas of action and their concrete design at a given time can be seen as the microfoundation of a phenomenon investigated at a later point in time (cf. Barney and Felin, 2013; Felin et al., 2012). However, a deeper engagement with the relevant literature is needed.

Finally, the use of Innovation Contests in other types of organizations, e.g., small and medium-sized enterprises or public institutions (such as universities) and the impact of the work environment as a determinant are almost neglected in this study.

---

[101] See also the Call for Papers by Felin et al. (2015) on "Organizing Crowds and Innovation".

References

Adamczyk, S., Bullinger, A.C. and Moeslein, K.M. (2010), "Call for attention – Attracting and activating innovators", *R&D Management Conference*.

Adamczyk, S., Bullinger, A.C. and Moeslein, K.M. (2011a), "Commenting for new ideas: insights from an open innovation platform", *International Journal of Technology Intelligence and Planning*, Vol. 7 No. 3, pp. 232–249.

Adamczyk, S., Bullinger, A.C. and Moeslein, K.M. (2012), "Innovation contests: A review, classification and outlook", *Creativity and Innovation Management*, Vol. 21 No. 4, pp. 335–360.

Adamczyk, S., Haller, J., Bullinger, A.C. and Moeslein, K.M. (2011b), "Knowing is silver, listening is gold: On the importance and impact of feedback in IT-based innovation contests", *Wirtschaftsinformatik Proceedings*.

Adamides, E.D. and Karacapilidis, N. (2006), "Information technology support for the knowledge and social processes of innovation management", *Technovation*, Vol. 26 No. 1, pp. 50–59.

Afuah, A. and Tucci, C.L. (2012), "Crowdsourcing as a solution to distant search", *Academy of Management Review*, Vol. 37 No. 3, pp. 355–375.

Ahimbisibwe, A., Cavana, R.Y. and Daellenbach, U.S. (2015), "A contingency fit model of critical success factors for software development projects: A comparison of agile and traditional plan-based methodologies", *Journal of Enterprise Information Management*, Vol. 28 No. 1, pp. 7–33.

Ahuja, M., Chudoba, K.M., George, J.F., Kacmar, C. and McKnight, H. (2002), "Overworked and isolated? Predicting the effect of work-family conflict, autonomy, and workload on organizational commitment and turnover of virtual Workers", *Proceedings of the Annual Hawaii International Conference on System Sciences*.

Aiken, L.S. and West, S.G. (1991), *Multiple regression: Testing and interpreting interactions,* 1st ed., Sage Publications, Thousand Oaks.

Ajzen, I. (1991), "The theory of planned behavior", *Organizational Behavior and Human Decision Processes*, Vol. 50 No. 2, pp. 179–211.

Alexandris, K., Kouthouris, C. and Girgolas, G. (2007), "Investigating the relationships among motivation, negotiation, and alpine skiing participation", *Journal of Leisure Research*, Vol. 39 No. 4, pp. 648–667.

Allen, N.J. and Meyer, J.P. (1990), "The measurement and antecedents of affective, continuance and normative commitment to the organization", *Journal of Occupational Psychology*, Vol. 63 No. 1, pp. 1–18.

Amabile, T.M. (1979), "Effects of external evaluation on artistic creativity", *Journal of Personality and Social Psychology*, Vol. 37 No. 2, pp. 221–233.

Amabile, T.M. (1983), "The social psychology of creativity: A componential conceptualization", *Journal of Personality and Social Psychology*, Vol. 45 No. 2, pp. 357–376.

Amabile, T.M. (1988), "A model of creativity and innovation in organizations", *Research in Organizational Behavior*, Vol. 10 No. 1, pp. 123–167.

Amabile, T.M. (1993), "Motivational synergy: Toward new conceptualizations of intrinsic and extrinsic motivation in the workplace", *Human Resource Management Review*, Vol. 3 No. 3, pp. 185–201.

Amabile, T.M. (1996), "Creativity and innovation in organizations", *Harvard Business School*.

Amabile, T.M. (1997), "Motivating creativity in organizations: On doing what you love and loving what you do", *California Management Review*, Vol. 40 No. 1, pp. 39–58.

Amabile, T.M. (2010), "KEYS sample feedback report", available at: http://www.ccl.org/leadership/assessments/KEYSOverview.aspx (accessed 26 May 2015).

Amabile, T.M., Burnside, R. and Gryskiewicz, S.S. (1995), "User's guide for KEYS: Assessing the climate for creativity".

Amabile, T.M., Conti, R., Coon, H., Lazenby, J. and Herron, M. (1996), "Assessing the work environment for creativity", *Academy of Management Journal*, Vol. 39 No. 5, pp. 1154–1184.

Amabile, T.M., Fisher, C.M. and Pillemer, J. (2014), "IDEO's culture of helping", *Harvard Business Review*, Vol. 92 No. 1, pp. 54–61.

Amabile, T.M. and Gryskiewicz, N.D. (1989), "The creative environment scales: Work environment inventory", *Creativity Research Journal*, Vol. 2 No. 4, pp. 231–253.

Amabile, T.M., Hill, K.G., Hennessey, B.A. and Tighe, E.M. (1994), "The work preference inventory: Assessing intrinsic and extrinsic motivational orientations", *Journal of Personality and Social Psychology*, Vol. 66 No. 5, pp. 950–967.

Amabile, T.M. and Khaire, M. (2008), "Creativity and the role of the leader", *Harvard Business Review*, Vol. 86 No. 10, pp. 100–109.

Amabile, T.M. and Kramer, S.J. (2007), "Inner work life: Understanding the subtext of business performance", *Harvard Business Review*, Vol. 85 No. 5, pp. 72–83.

Amabile, T.M. and Pillemer, J. (2012), "Perspectives on the social psychology of creativity", *The Journal of Creative Behavior*, Vol. 46 No. 1, pp. 3–15.

Amabile, T.M., Schatzel, E.A., Moneta, G.B. and Kramer, S.J. (2004), "Leader behaviors and the work environment for creativity: Perceived leader support", *The Leadership Quarterly*, Vol. 15 No. 1, pp. 5–32.

Anderson, J.C. and Gerbing, D.W. (1984), "The effect of sampling error on convergence, improper solutions, and goodness-of-fit indices for maximum likelihood confirmatory factor analysis", *Psychometrika*, Vol. 49 No. 2, pp. 155–173.

Anderson, J.C. and Gerbing, D.W. (1988), "Structural equation modeling in practice: A review and recommended two-step approach", *Psychological Bulletin*, Vol. 103 No. 3, pp. 411–423.

Argyres, N.S. and Silverman, B.S. (2004), "R&D, organization structure, and the development of corporate technological knowledge", *Strategic Management Journal*, Vol. 25 No. 89, pp. 929–958.

Armisen, A. and Majchrzak, A. (2015), "Tapping the innovative business potential of innovation contests", *Business Horizons*, Vol. 58 No. 1, pp. 389–399.

Aselage, J. and Eisenberger, R. (2003), "Perceived organizational support and psychological contracts: A theoretical integration", *Journal of Organizational Behavior*, Vol. 24 No. 5, pp. 491–509.

Baregheh, A., Rowley, J. and Sambrook, S. (2009), "Towards a multidisciplinary definition of innovation", *Management Decision*, Vol. 47 No. 8, pp. 1323–1339.

Barney, J. and Felin, T. (2013), "What are microfoundations?", *Academy of Management Perspectives*, Vol. 27 No. 2, pp. 138–155.

Baron and Kenny (1986), "The moderator-mediator variable distinction in social psychological research: Conceptual, strategic, and statistical considerations", *Journal of Personality and Social Psychology*, Vol. 51 No. 6, pp. 1173–1182.

Bateman, P.J., Gray, P.H. and Butler, B.S. (2011), "Research Note - The impact of community commitment on participation in online communities", *Information Systems Research*, Vol. 22 No. 4, pp. 841–854.

Bayus, B.L. (2013), "Crowdsourcing new product ideas over time: An analysis of the Dell IdeaStorm community", *Management Science*, Vol. 59 No. 1, pp. 226–244.

Becker, T.E. (2005), "Potential problems in the statistical control of variables in organizational research: A qualitative analysis with recommendations", *Organizational Research Methods*, Vol. 8 No. 3, pp. 274–289.

Bertaux, D. (1981), "From the life-history approach to the transformation of sociological practice", in Bertaux, D. (Ed.), *Biography and society: The life history approach in the social sciences*, 1st ed., Sage Publications, London, pp. 29–45.

Bjelland, O.M. and Wood, R.C. (2008), "An inside view of IBM's' Innovation Jam'", *MIT Sloan Management Review*, Vol. 50 No. 1, pp. 32–40.

Blau, P.M. (1964), *Exchange and power in social life*, 1st ed., Transaction Publishers, New Brunswick.

Blohm, I., Bretschneider, U., Leimeister, J.M. and Krcmar, H. (2011a), "Does collaboration among participants lead to better ideas in IT-based idea competitions? An empirical investigation", *International Journal of Networking and Virtual Organisations*, Vol. 9 No. 2, pp. 106–122.

Blohm, I., Leimeister, J.M., Rieger, M. and Krcmar, H. (2011b), "Controlling von Ideencommunities – Entwicklung und Test einer Ideencommunity-Scorecard", *Controlling*, Vol. 23 No. 2, pp. 96–103.

Blohm, I., Riedl, C., Leimeister, J.M. and Krcmar, H. (2011c), "Idea evaluation mechanisms for collective intelligence in open innovation communities: Do traders outperform raters?", *Proceedings of the International Conference on Information Systems*.

BMBF (2015), "Announcement on the research focus 'work in the digital world'", available at: https://www.bmbf.de/foerderungen/bekanntmachung.php?B=1017 (accessed 21 May 2015).

Bolton, P. and Dewatripont, M. (1994), "The firm as a communication network", *The Quarterly Journal of Economics*, Vol. 109 No. 4, pp. 809–839.

Borowiak, Y. and Herrmann, T. (2011), "Web 2.0 zur Unterstützung von Innovationsarbeit", in Howaldt, J., Kopp, R. and Beerheide, E. (Eds.), *Innovationsmanagement 2.0: Handlungsorientierte Einführung und praxisbasierte Impulse*, 1st ed., Gabler, Wiesbaden, pp. 131–154.

Boudreau, K.J., Lacetera, N. and Lakhani, K.R. (2011), "Incentives and problem uncertainty in innovation contests: An empirical analysis", *Management Science*, Vol. 57 No. 5, pp. 843–863.

Bryman, A. (2012), *Social research methods*, 4th ed., Oxford University Press, Oxford.

Bullinger, A. and Moeslein, K.M. (2010), "Innovation contests - Where are we?", *Proceedings of the Americas Conference on Information Systems*.

Bullinger, A.C., Neyer, A.-K., Rass, M. and Moeslein, K.M. (2010), "Community-based innovation contests: Where competition meets cooperation", *Creativity and Innovation Management*, Vol. 19 No. 3, pp. 290–303.

Cahalane, M., Finnegan, P. and Feller, J. (2013), "Peer produced innovation: An exploration of 'the wisdom of crowds' in virtual worlds", *Proceedings of the European Conference on Information Systems*.

Chen, J., Xu, H. and Whinston, A.B. (2011), "Moderated online communities and quality of user-generated content", *Journal of Management Information Systems*, Vol. 28 No. 2, pp. 237–268.

Chen, Z., Eisenberger, R., Johnson, K.M., Sucharski, I.L. and Aselage, J. (2009), "Perceived organizational support and extra-role performance: Which leads to which?", *The Journal of Social Psychology*, Vol. 149 No. 1, pp. 119–124.

Chesbrough, H. and Brunswicker, S. (2014), "A fad or a phenomenon?: The adoption of open innovation practices in large firms", *Research-Technology Management*, Vol. 57 No. 2, pp. 16–25.

Chiaroni, D., Chiesa, V. and Frattini, F. (2011), "The open innovation journey: How firms dynamically implement the emerging innovation management paradigm", *Technovation*, Vol. 31 No. 1, pp. 34–43.

Chin, W.W. (1998), "Commentary: Issues and opinion on structural equation modeling", *MIS Quarterly*, Vol. 22 No. 1, pp. vii–xvi.

Churchill, G.A. (1979), "A paradigm for developing better measures of marketing constructs", *Journal of Marketing Research*, Vol. 16 No. 1, pp. 64–73.

Coates, T.T. and McDermott, C.M. (2002), "An exploratory analysis of new competencies: A resource based view perspective", *Journal of Operations Management*, Vol. 20 No. 5, pp. 435–450.

Cohen, C., Kaplan, T.R. and Sela, A. (2008), "Optimal rewards in contests", *The RAND Journal of Economics*, Vol. 39 No. 2, pp. 434–451.

Collis, D.J. (1994), "Research note: How valuable are organizational capabilities?", *Strategic Management Journal*, Vol. 15 Special Issue: Competitive Organizational Behavior, pp. 143–152.

Coltman, T., Devinney, T.M., Midgley, D.F. and Venaik, S. (2008), "Formative versus reflective measurement models: Two applications of formative measurement", *Journal of Business Research*, Vol. 61 No. 12, pp. 1250–1262.

Cooper, R.G. and Kleinschmidt, E.J. (1995), "Benchmarking the firm's critical success factors in new product development", *Journal of Product Innovation Management*, Vol. 12 No. 5, pp. 374–391.

Cooper, R.G. and Kleinschmidt, E.J. (2007), "Winning businesses in product development: The critical success factors", *Research-Technology Management*, Vol. 50 No. 3, pp. 52–66.

Costello, A.B. and Osborne, J. (2005), "Best practices in exploratory factor analysis: four recommendations for getting the most from your analysis", *Practical Assessment Research & Evaluation*, Vol. 10 No. 7, pp. 1–9.

Creswell, J.W. (1998), *Qualitative inquiry and research design: Choosing among five traditions*, 1st ed., Sage Publications, Thousand Oaks.

Creswell, J.W. (2013), *Research design: Qualitative, quantitative, and mixed methods approaches*, 2nd ed., Sage Publications, Thousand Oaks.

Cropanzano, R. (2005), "Social exchange theory: An interdisciplinary review", *Journal of Management*, Vol. 31 No. 6, pp. 874–900.

Cropanzano, R., Howes, J.C., Grandey, A.A. and Toth, P. (1997), "The relationship of organizational politics and support to work behaviors, attitudes, and stress", *Journal of Organizational Behavior*, Vol. 18 No. 2, pp. 159–180.

Crossan, M.M. and Apaydin, M. (2010), "A multi-dimensional framework of organizational innovation: A systematic review of the literature", *Journal of Management Studies*, Vol. 47 No. 6, pp. 1154–1191.

Dahlander, L. and Gann, D.M. (2010), "How open is innovation?", *Research Policy*, Vol. 39 No. 6, pp. 699–709.

De Brentani, U. and Reid, S.E. (2012), "The fuzzy front-end of discontinuous innovation: Insights for research and management", *Journal of Product Innovation Management*, Vol. 29 No. 1, pp. 70–87.

Demerouti, E., Bakker, A.B., Nachreiner, F. and Schaufeli, W.B. (2001), "The job demands-resources model of burnout", *Journal of Applied Psychology*, Vol. 86 No. 3, pp. 499–512.

Desouza, K.C., Dombrowski, C., Awazu, Y., Baloh, P., Papagari, S., Jha, S. and Kim, J.Y. (2009), "Crafting organizational innovation processes", *Innovation*, Vol. 11 No. 1, pp. 6–33.

Di Gangi, P. M., Wasko, M. and Hooker, R. (2010), "Getting customers' ideas to work for you: Learning from Dell how to succeed with online user innovation communities", *MIS Quarterly Executive*, Vol. 9 No. 4, pp. 213–228.

Diener, K. and Piller, F.T. (2013), *The market for open innovation: Increasing the efficiency and effectiveness of the innovation process*, 2nd ed., Lulu Publishing, Raleigh.

Digmayer, C. and Jakobs, E.-M. (2013), "Shared Ideas: Integration von Open-Innovation-Plattform-Methoden in Design-Thinking-Prozesse", in Keuper, F., Hamidian, K., Verwaayen, E., Kalinowski, T. and Kraijo, C. (Eds.), *Digitalisierung und Innovation: Planung - Entstehung - Entwicklungsperspektiven*, 1st ed., Springer Gabler, Wiesbaden, pp. 365–394.

Dresing, T., Pehl, T. and Schmieder, C. (2015), "Manual (on) transcription. Transcription conventions, foftware guides and practical hints for qualitative researchers", available at: http://www.audiotranskription.de/english/transcription-practicalguide.htm (accessed 29 June 2015).

Duke, A.B., Goodman, J.M., Treadway, D.C. and Breland, J.W. (2009), "Perceived organizational support as a moderator of emotional labor/outcomes relationships", *Journal of Applied Social Psychology*, Vol. 39 No. 5, pp. 1013–1034.

Dumbach, M. (2014), *Establishing corporate innovation communities: A social capital perspective*, Springer, Wiesbaden.

Duriau, V.J., Reger, R.K. and Pfarrer, M.D. (2007), "A content analysis of the content analysis literature in organization studies: Research themes, data sources, and methodological refinements", *Organizational Research Methods*, Vol. 10 No. 1, pp. 5–34.

Dutta, S., Narasimhan, O. and Rajiv, S. (2005), "Conceptualizing and measuring capabilities: Methodology and empirical application", *Strategic Management Journal*, Vol. 26 No. 3, pp. 277–285.

Dziuban, C.D. and Shirkey, E.C. (1974), "When is a correlation matrix appropriate for factor analysis? Some decision rules.", *Psychological Bulletin*, Vol. 81 No. 6, p. 358.

Ebner, W., Leimeister, J.M. and Krcmar, H. (2009), "Community engineering for innovations: The ideas competition as a method to nurture a virtual community for innovations", *R&D Management*, Vol. 39 No. 4, pp. 342–356.

Eisenberger, R., Armeli, S., Rexwinkel, B., Lynch, P.D. and Rhoades, L. (2001), "Reciprocation of perceived organizational support", *Journal of Applied Psychology*, Vol. 86 No. 1, pp. 42–51.

Eisenberger, R., Cotterell, N. and Marvel, J. (1987), "Reciprocation ideology", *Journal of Personality and Social Psychology*, Vol. 53 No. 4, pp. 743–750.

Eisenberger, R., Fasolo, P. and Davis-LaMastro, V. (1990), "Perceived organizational support and employee diligence, commitment, and innovation", *Journal of Applied Psychology*, Vol. 75 No. 1, pp. 51–59.

Eisenberger, R., Huntington, R., Hutchison, S. and Sowa, D. (1986), "Perceived organizational support", *Journal of Applied Psychology*, Vol. 71 No. 3, pp. 500–507.

Eisenberger, R., Karagonlar, G., Stinglhamber, F., Neves, P., Becker, T.E., Gonzalez-Morales, M.G. and Steiger-Mueller, M. (2010), "Leader-member exchange and affective organizational commitment: The contribution of supervisor's organizational embodiment", *Journal of Applied Psychology*, Vol. 95 No. 6, pp. 1085–1103.

Eisenberger, R. and Shanock, L.R. (2003), "Rewards, intrinsic motivation, and creativity: A case study of conceptual and methodological isolation", *Creativity Research Journal*, Vol. 15 No. 2, pp. 121–130.

Eisenberger, R., Stinglhamber, F., Vandenberghe, C., Sucharski, I.L. and Rhoades, L. (2002), "Perceived supervisor support: Contributions to perceived organizational support and employee retention", *Journal of Applied Psychology*, Vol. 87 No. 3, pp. 565–573.

Elaine Botha, M. (1989), "Theory development in perspective: The role of conceptual frameworks and models in theory development", *Journal of advanced nursing*, Vol. 14 No. 1, pp. 49–55.

Erdogan, B. and Enders, J. (2007), "Support from the top: Supervisors' perceived organizational support as a moderator of leader-member exchange to satisfaction and performance relationships", *The Journal of applied psychology*, Vol. 92 No. 2, pp. 321–330.

Erickson, L.B., Trauth, E.M. and Petrick, I. (2012), "Getting inside your employees' heads: Navigating barriers to internal-crowdsourcing for product and service innovation", *Proceedings of the International Conference on Information Systems*.

Farrell, A.M. and Rudd, J.M. (2009), "Factor analysis and discriminant validity: A brief review of some practical issues", *Australian & New Zealand Marketing Academy Conference*.

Felin, T., Foss, N.J., Heimeriks, K.H. and Madsen, T.L. (2012), "Microfoundations of routines and capabilities: Individuals, processes, and structure", *Journal of Management Studies*, Vol. 49 No. 8, pp. 1351–1374.

Felin, T., Lakhani, K.R. and Tushman, M.L. (2015), "Special issue of strategic organization: "Organizing crowds and innovation"", *Strategic Organization*, Vol. 13 No. 1, pp. 86–87.

Fornell, C. and Larcker, D.F. (1981), "Evaluating structural equation models with unobservable variables and measurement error", *Journal of Marketing Research*, Vol. 18 No. 1, pp. 39–50.

Foss, N.J., Laursen, K. and Pedersen, T. (2011), "Linking customer interaction and innovation: The mediating role of new organizational practices", *Organization Science*, Vol. 22 No. 4, pp. 980–999.

Franke, N., Keinz, P. and Klausberger, K. (2013), "Does this sound like a fair deal? Antecedents and consequences of fairness expectations in the individual's decision to participate in firm innovation", *Organization Science*, Vol. 24 No. 5, pp. 1495–1516.

Franke, N., Lettl, C., Roiser, S. and Tuertscher, P. (2014), ""Does god play dice?"- Randomness vs. deterministic explanations of crowdsourcing success", *Academy of Management Proceedings*.

Franke, N. and Shah, S. (2003), "How communities support innovative activities: An exploration of assistance and sharing among end-users", *Research Policy*, Vol. 32 No. 1, pp. 157–178.

Frey, K., Luethje, C. and Haag, S. (2011), "Whom should firms attract to open innovation platforms? The role of knowledge diversity and motivation", *Long Range Planning*, Vol. 44 No. 5, pp. 397–420.

Fueller, J., Moeslein, K.M., Hutter, K. and Haller, J. (2010), "Evaluation games - How to make the crowd your jury", *GI-Jahrestagung*.

Gable, G.G. (1994), "Integrating case study and survey research methods: An example in information systems", *European Journal of Information Systems*, Vol. 3 No. 2, pp. 112–126.

Gandz, J. and Murray, V.V. (1980), "The experience of workplace politics", *Academy of Management Journal*, Vol. 23 No. 2, pp. 237–251.

Gebauer, J., Fueller, J. and Pezzei, R. (2013), "the dark and the bright side of co-creation: Triggers of member behavior in online innovation communities", *Journal of Business Research*, Vol. 66, pp. 1516–1527.

Gefen, D., Straub, D.W. and Boudreau, M.-C. (2000), "Structural equation modeling and regression: Guidelines for research practice", *Communicartions of the AIS*, Vol. 4 No. 1, pp. 1–77.

Gollwitzer, P.M. (1990), "Action phases and mind-set", in Higgins, E. and Sorrentino, R.M. (Eds.), *Handbook of motivation and cognition: Foundations of social behavior*, 2nd ed., Guilford Press, New York, pp. 53–92.

Gollwitzer, P.M., Heckhausen, H. and Ratajczak, H. (1990), "From weighing to willing: Approaching a change decision through pre- or postdecisional mentation", *Organizational Behavior and Human Decision Processes*, Vol. 45 No. 1, pp. 41–65.

Gorgievski, M.J. and Hobfoll, S.E. (2008), "Work can burn us out or fire us up: Conservation of resources in burnout and engagement", in Halbesleben, J.R.B. (Ed.), *Handbook of Stress and Burnout in Health Care*, 1st ed., Nova Science Publishers, pp. 7–22.

Govindarajan, V. (2011), "Innovation's nine critical success factors", available at: https://hbr.org/2011/07/innovations-9-critical-success/ (accessed 5 March 2016).

Grant, R.M. (1996), "Toward a knowledge-based theory of the firm", *Strategic Management Journal*, Vol. 17 Special Issue: Knowledge and the Firm, pp. 109–122.

Grote, M., Herstatt, C. and Gemuenden, H.G. (2012), "Cross-divisional innovation in the large corporation: Thoughts and evidence on its value and the role of the early stages of innovation", *Creativity and Innovation Management*, Vol. 21 No. 4, pp. 361–375.

Guest, G., Bunce, A. and Johnson, L. (2006), "How many interviews are enough? An experiment with data saturation and variability", *Field Methods*, Vol. 18 No. 1, pp. 59–82.

Halbesleben, J.R.B. and Bowler, W.M. (2007), "Emotional exhaustion and job performance: The mediating role of motivation", *The Journal of applied psychology*, Vol. 92 No. 1, pp. 93–106.

Haller, J.B.A., Bullinger, A.C. and Moeslein, K.M. (2011), "Innovation Contests. An IT-based tool for innovation management", *Business & Information Systems Engineering*, Vol. 3 No. 2, pp. 103–106.

Hallerstede, S.H. and Bullinger, A.C. (2010), "Do you know where you go - A taxonomy of online innovation contests", *Proceedings of the ISPIM Conference*.

Hamel, G. (1991), "Competition for competence and inter-partner learning within international strategic alliances", *Strategic Management Journal*, Vol. 12 Special Issue: Global Strategy, pp. 83–103.

Hayes, A.F. (2013), *Introduction to mediation, moderation, and conditional process analysis: A regression-based approach,* 1st ed., Guilford Press, New York.

Heckhausen, H. and Gollwitzer, P.M. (Eds.) (1986), *Information processing before and after the formation of an intent*, Elsevier, Amsterdam.

Heckhausen, H. and Gollwitzer, P.M. (1987), "Thought contents and cognitive functioning in motivational versus volitional states of mind", *Motivation and emotion*, Vol. 11 No. 2, pp. 101–120.

Henseler, J., Ringle, C.M. and Sarstedt, M. (2015), "A new criterion for assessing discriminant validity in variance-based structural equation modeling", *Journal of the Academy of Marketing Science*, Vol. 43 No. 1, pp. 115–135.

Hetmank, L. (2013), "Towards a semantic standard for enterprise crowdsourcing - A scenario-based evaluation of a conceptual prototype", *Proceedings of the European Conference on Information Systems*.

Hidalgo, A. and Albors, J. (2008), "Innovation management techniques and tools: A review from theory and practice", *R&D Management*, Vol. 38 No. 2, pp. 113–127.

Hoeber, B. (2015), "Innovation contests within firm boundaries – Areas of action to increase organizational support", *The Annual Open and User Innovation Society Meeting*.

Hoeber, B., Schaarschmidt, M. and Von Kortzfleisch, H. (2016), "Work environment perceptions as determinants of affective commitment and participation in firm-internal innovation contests", *Proceedings of the European Conference on Information Systems*.

Hooper, D., Coughlan, J. and Mullen, M. (2008), "Structural equation modelling: Guidelines for determining model fit", *Electronic Journal of Business Research Methods*, Vol. 6 No. 1, pp. 53–60.

Houkes, I., Janssen, P.P., Jonge, J. de and Bakker, A.B. (2003), "Specific determinants of intrinsic work motivation, emotional exhaustion and turnover intention: A multisample longitudinal study", *Journal of Occupational and Organizational Psychology*, Vol. 76 No. 4, pp. 427–450.

Huber, G.P. (1990), "A theory of the effects of advanced information technologies on organizational design, intelligence, and decision making", *Academy of Management Review*, Vol. 15 No. 1, pp. 47–71.

Hutter, K., Hautz, J., Fueller, J., Mueller, J. and Matzler, K. (2011), "Communitition: The tension between competition and collaboration in community-based design contests", *Creativity and Innovation Management*, Vol. 20 No. 1, pp. 3–21.

Iacobucci, D. (2010), "Structural equations modeling: Fit Indices, sample size, and advanced topics", *Journal of Consumer Psychology*, Vol. 20 No. 1, pp. 90–98.

Ihl, C., Piller, F.T. and Wagner, P. (2012a), "Organizing for open innovation: Aligning internal structure with external knowledge search", *SSRN Working paper*, No. 2164766.

Ihl, C., Vossen, A. and Piller, F.T. (2012b), "All for the money? The ambiguity of monetary rewards in firm-initiated ideation with users", *SSRN Working paper*, No. 2164763.

Jarvenpaa, S.L. and Ives, B. (1991), "Executive involvement and participation in the management of information technology", *MIS Quarterly*, Vol. 15 No. 2, pp. 205–227.

Jeppesen, L.B. and Frederiksen, L. (2006), "Why do users contribute to firm-hosted user communities? The case of computer-controlled music instruments", *Organization Science*, Vol. 17 No. 1, pp. 45–63.

Johnson, R.B., Onwuegbuzie, A.J. and Turner, L.A. (2007), "Toward a definition of mixed methods research", *Journal of Mixed Methods Research*, Vol. 1 No. 2, pp. 112–133.

Jouret, G. (2009), "Inside Cisco's search for the next big idea", *Harvard Business Review*, Vol. 87 No. 9, pp. 43–45.

Jung, Y., Majchrzak, A., Malhotra, A. and Johnson, J. (2012), "Encouraging collaborative idea-building in enterprise-wide innovation challenges", *Proceedings of the International Conference on Information Systems*.

Kaiser, H.F. and Rice, J. (1974), "Little Jiffy, Mark IV", *Educational and Psychological Measurement*, Vol. 34 No. 1, pp. 111–117.

Kaplan, A.M. and Haenlein, M. (2010), "Users of the world, unite! The challenges and opportunities of Social Media", *Business Horizons*, Vol. 53 No. 1, pp. 59–68.

Katila, R. and Ahuja, G. (2002), "Something old, something new: A longitudinal study of search behavior and new product introduction", *Academy of Management Journal*, Vol. 45 No. 6, pp. 1183–1194.

Keuper, F., Hamidian, K., Verwaayen, E., Kalinowski, T. and Kraijo, C. (Eds.) (2013), *Digitalisierung und Innovation: Planung - Entstehung - Entwicklungsperspektiven*, 1st ed., Springer Gabler, Wiesbaden.

Khurana, A. and Rosenthal, S.R. (1998), "Towards holistic "front ends" in new product development", *Journal of Product Innovation Management*, Vol. 15 No. 1, pp. 57–74.

Kim, D.J., Ferrin, D.L. and Rao, H.R. (2008), "A trust-based consumer decision-making model in electronic commerce: The role of trust, perceived risk, and their antecedents", *Decision Support Systems*, Vol. 44 No. 2, pp. 544–564.

Kim, J. and Wilemon, D. (2002), "Focusing the fuzzy front–end in new product development", *R&D Management*, Vol. 32 No. 4, pp. 269–279.

Kim, K.Y., Eisenberger, R. and Baik, K. (2016), "Perceived organizational support and affective organizational commitment: Moderating influence of perceived organizational competence", *Journal of Organizational Behavior*, Vol. 37 No. 4, pp. 558–583.

Kimberley, J.R. and Evanisko, M.J. (1981), "Organizational innovation: The influence of individual, organizational, and contextual factors on hospital adoption of technological and administrative innovations", *The Academy of Management Journal*, Vol. 24 No. 4, pp. 689–713.

King, R. (2014), "AT&T develops employee ideas for innovation", available at: http://blogs.wsj.com/cio/2014/11/12/att-develops-employee-ideas-for-innovation/ (accessed 18 May 2016).

Kivimaeki, M., Laensisalmi, H., Elovainio, M., Heikkilae, A. and Lindstroem, K. (2000), "Communication as a determinant of organizational innovation", *R&D Management*, Vol. 30 No. 1, pp. 33–42.

Klein, H.K. and Myers, M.D. (1999), "A set of principles for conducting and evaluating interpretive field studies in information systems", *MIS Quarterly*, Vol. 23 No. 1, pp. 67–93.

Kline, R.B. (2011), *Principles and practice of structural equation modeling*, 3rd ed., Guilford Press, New York, London.

Koen, P.A., Ajamian, G.M., Boyce, S., Clamen, A., Fisher, E., Fountoulakis, S., Johnson, A., Puri, P. and Seibert, R. (2004), "Fuzzy front end: Effective methods, tools, and techniques", in Belliveau, P., Griffin, A. and Somermeyer, S. (Eds.), *The PDMA ToolBook for New Product Development*, 1st ed., Wiley, New York, pp. 5–35.

Koh, J., Kim, Y.-G., Butler, B. and Bock, G.-W. (2007), "Encouraging participation in virtual communities", *Communications of the ACM*, Vol. 50 No. 2, pp. 68–73.

Kopp, R. (2011), "Enterprise 2.0 als soziodigitales Innovationssystem", in Howaldt, J., Kopp, R. and Beerheide, E. (Eds.), *Innovationsmanagement 2.0: Handlungsorientierte Einführung und praxisbasierte Impulse*, 1st ed., Gabler, Wiesbaden, pp. 37–65.

Krippendorff, K. (2004), *Content analysis: An introduction to its methodology*, 2nd edition, Sage Publications, Thousand Oaks.

Kubátová, J. (2012), "Innovative managerial principles for current knowledge economy", *Economics and Management*, Vol. 17 No. 1, pp. 359–364.

Kuckartz, U. (2011), *Mixed Methods: Methodologie, Forschungsdesigns und Analyseverfahren*, 1st ed., Springer VS, Wiesbaden.

Kumar, N., Stern, L.W. and Anderson, J.C. (1993), "Conducting interorganizational research using key informants", *Academy of Management Journal*, Vol. 36 No. 6, pp. 1633–1651.

Kuratko, D.F., Montagno, R.V. and Hornsby, J.S. (1990), "Developing an intrapreneurial assessment instrument for an effective corporate entrepreneurial environment", *Strategic Management Journal*, Vol. 11, pp. 49–58.

Kurtessis, J.N., Eisenberger, R., Ford, M.T., Buffardi, L.C., Stewart, K.A. and Adis, C.S. (2015), "Perceived organizational support: A meta-analytic evaluation of organizational support theory", *Journal of Management (in press)*.

Kurtmollaiev, S. (2012), "Open innovation and NIH syndrome through the lens of transaction cost", *Proceedings of the DRUID Conference*.

Lawson, B. and Samson, D. (2001), "Developing innovation capability in organisations: A dynamic capabilities approach", *International Journal of Innovation Management*, Vol. 5 No. 3, pp. 377–400.

Lee, R.T. and Ashforth, B.E. (1996), "A meta-analytic examination of the correlates of the three dimensions of job burnout", *Journal of Applied Psychology*, Vol. 81 No. 2, pp. 123–133.

Leimeister, J.M., Huber, M., Bretschneider, U. and Krcmar, H. (2009), "Leveraging crowdsourcing: Activation-supporting components for IT-based ideas competition", *Journal of Management Information Systems*, Vol. 26 No. 1, pp. 197–224.

LePine, J.A., Podsakoff, N.P. and LePine, M.A. (2005), "A meta-analytic test of the challenge stressor–hindrance stressor framework: An explanation for inconsistent relationships among stressors and performance", *Academy of Management Journal*, Vol. 48 No. 5, pp. 764–775.

Liedtka, J. (2011), "Learning to use design thinking tools for successful innovation", *Strategy & Leadership*, Vol. 39 No. 5, pp. 13–19.

Lin, T.Y., Chuang, L.M., Chang, M.Y. and Yeh, C.M. (2012), "A study of the relationship between team innovation and organizational innovation in high tech industry: Confirmation of the organizational culture moderation effect", *Advances in Management & Applied Economics*, Vol. 2 No. 2, pp. 19–52.

Luettgens, D., Pollok, P., Antons, D. and Piller, F. (2014), "Wisdom of the crowd and capabilities of a few: Internal success factors of crowdsourcing for innovation", *Journal of Business Economics*, Vol. 84 No. 3, pp. 339–374.

Ma, M. and Agarwal, R. (2007), "Through a glass darkly: Information technology design, identity verification, and knowledge contribution in online communities", *Information Systems Research*, Vol. 18 No. 1, pp. 42–67.

Machlup, F. (1967), "Theories of the firm: Marginalist, behavioral, managerial", *American Economic Review*, Vol. 57 No. 1, pp. 201–220.

Maimbo, H. and Pervan, G. (2005), "Designing a case study protocol for application in IS research", *Proceedings of the Pacific Asia Conference on Information Systems*.

MarketsandMarkets (2013), *Market Research Report: Global Enterprise Social Software (ESS) Market: Global Advancements, Demand Analysis & Worldwide Market Forecasts (2013-2018)*.

Marshall, M.N. (1996), "The key informant technique", *Family Practice*, Vol. 13 No. 1, pp. 92–97.

Mathieu, J.E. and Zajac, D.M. (1990), "A review and meta-analysis of the antecedents, correlates, and consequences of organizational commitment", *Psychological Bulletin*, Vol. 108 No. 2, pp. 171–194.

Mayer, K.J. and Sparrowe, R.T. (2013), "Integrating theories in AMJ articles", *Academy of Management Journal*, Vol. 56 No. 4, pp. 917–922.

Mayring, P. (2004), "Qualitative content analysis", *Forum: Qualitative Social Research*.

Mayring, P. (2014), "Qualitative content analysis: Theoretical foundation, basic procedures and software solution", available at: http://nbn-resolving.de/urn:nbn:de:0168-ssoar-395173 (accessed 20 February 2015).

McAfee, A.P. (2006), "Enterprise 2.0: The dawn of emergent collaboration", *MIT Sloan Management Review*, Vol. 47 No. 3, pp. 21–28.

McAfee, A.P. (2009), "Enterprise 2.0: How a connected workforce innovates", *Harvard Business Review*, Vol. 87 No. 12, p. 80.

McAfee, A.P. (2014), "Enterprise 2.0, Version 2.0", available at: http://andrewmcafee.org/2006/05/enterprise_20_version_20/ (accessed 1 September 2014).

McGourty, J., Tarshis, L.A. and Dominick, P. (1996), "Managing innovation: Lessons from world class organizations", *International Journal of Technology Management*, Vol. 11 No. 3, pp. 354–368.

McLean, L.D. (2005), "Organizational culture's influence on creativity and innovation: A review of the literature and implications for human resource development", *Advances in Developing Human Resources*, Vol. 7 No. 2, pp. 226–246.

McNally, R.C., Akdeniz, M.B. and Calantone, R.J. (2011), "New product development processes and new product profitability: Exploring the mediating role of speed to market and product quality", *Journal of Product Innovation Management*, Vol. 28 No. S1, pp. 63–77.

Meisterjahn, M. (2015), "Inbound Innovation Management", available at: http://blog.hypeinnovation.com/inbound-innovation-management (accessed 5 October 2015).

Meyer, J.P. and Allen, N.J. (1991), "A three-component conceptualization of organizational commitment", *Human Resource Management Review*, Vol. 1 No. 1, pp. 61–89.

Meyer, J.P., Becker, T.E. and Vandenberghe, C. (2004), "Employee commitment and motivation: A conceptual analysis and integrative model", *Journal of Applied Psychology*, Vol. 89 No. 6, pp. 991–1007.

Meyer, J.P., Stanley, D.J., Herscovitch, L. and Topolnytsky, L. (2002), "Affective, continuance, and normative commitment to the organization: A meta-analysis of antecedents, correlates, and consequences", *Journal of Vocational Behavior*, Vol. 61 No. 1, pp. 20–52.

Miles, M.B. and Huberman, A.M. (1984), "Drawing valid meaning from qualitative data: Toward a shared craft", *Educational Researcher*, Vol. 13 No. 5, pp. 20–30.

Miles, M.B., Huberman, A.M. and Saldaña, J. (2013), *Qualitative Data Analysis: A Methods Sourcebook*, 3rd ed., Sage Publications, Thousand Oaks.

Moenaert, R., Caeldries, F., Lievens, A. and Wauters, E. (2000), "Communication flows in international product innovation teams", *Journal of Product Innovation Management*, Vol. 17 No. 5, pp. 360–377.

Moeslein, K.M. (2013), "Open innovation: Actors, tools, and tensions", in Huff, A.S., Moeslein, K.M. and Reichwald, R. (Eds.), *Leading open innovation*, 1st ed., MIT Press, Cambridge.

Mudambi, R., Mudambi, S.M. and Navarra, P. (2007), "Global innovation in MNCs: The effects of subsidiary self-determination and teamwork*", *Journal of Product Innovation Management*, Vol. 24 No. 5, pp. 442–455.

Mudambi, R. and Navarra, P. (2004), "Is knowledge power? Knowledge flows, subsidiary power and rent-seeking within MNCs", *Journal of International Business Studies*, Vol. 35 No. 5, pp. 385–406.

Neves, P. and Eisenberger, R. (2012), "Management Communication and Employee Performance: The Contribution of Perceived Organizational Support", *Human Performance*, Vol. 25 No. 5, pp. 452–464.

Neyer, A.-K., Bullinger, A.C. and Moeslein, K.M. (2009), "Integrating inside and outside innovators: A sociotechnical systems perspective", *R&D Management*, Vol. 39 No. 4, pp. 410–419.

Nobelius, D. (2004), "Towards the sixth generation of R&D management", *International Journal of Project Management*, Vol. 22 No. 5, pp. 369–375.

Obstfeld, D. (2005), "Social networks, the tertius iungens orientation, and involvement in innovation", *Administrative Science Quarterly*, Vol. 50 No. 1, pp. 100–130.

Odih, P. and Knights, D. (2002), "Now's the time! Consumption and time (space) disruptions in postmodern virtual worlds", in Whipp, R., Adam, B. and Sabelis, I. (Eds.), *Making time: Time and management in modern organizations*, 1st ed., Oxford University Press, Oxford, pp. 61–76.

Oliveira, M., Bitencourt, C., Teixeira, E. and Santos, A.C. (2013), "Thematic content analysis: Is there a difference between the support provided by the MAXQDA® and NVivo® software packages", *Proceedings of the European Conference on Research Methods for Business and Management Studies*.

Ortt, J.R. and Smits, R. (2006), "Innovation management: Different approaches to cope with the same trends", *International Journal of Technology Management*, Vol. 34 No. 3-4, pp. 296–318.

Pierce, J.L., Gardner, D.G., Cummings, L.L. and Dunham, R.B. (1989), "Organization-based self-esteem: Construct definition, measurement, and validation", *Academy of Management Journal*, Vol. 32 No. 3, pp. 622–648.

Piller, F., Vossen, A. and Ihl, C. (2012), "From social media to social product development: The impact of social media on co-Creation of innovation", *Die Unternehmung*, Vol. 65 No. 1, pp. 1–22.

Piller, F.T. and Walcher, D. (2006), "Toolkits for idea competitions: A novel method to integrate users in new product development", *R&D Management*, Vol. 36 No. 3, pp. 307–318.

Pinder, C.C. (2014), *Work motivation in organizational behavior*, 2nd ed., Psychology Press, New York.

Pirola-Merlo, A. and Mann, L. (2004), "The relationship between individual creativity and team creativity: Aggregating across people and time", *Journal of Organizational Behavior*, Vol. 25 No. 2, pp. 235–257.

Podsakoff, P.M., MacKenzie, S.B., Lee, J.-Y. and Podsakoff, N.P. (2003), "Common method biases in behavioral research: A critical review of the literature and recommended remedies", *Journal of Applied Psychology*, Vol. 88 No. 5, pp. 879–903.

Poetz, M.K. and Schreier, M. (2012), "The Value of crowdsourcing: Can users really compete with professionals in generating new product ideas?", *Journal of Product Innovation Management*, Vol. 29 No. 2, pp. 245–256.

Preacher, K.J. and Hayes, A.F. (2004), "SPSS and SAS procedures for estimating indirect effects in simple mediation models", *Behavior Research Methods, Instruments, & Computers*, Vol. 36 No. 4, pp. 717–731.

Presser, S., Couper, M.P., Lessler, J.T., Martin, E., Martin, J., Rothgeb, J.M. and Singer, E. (2004), "Methods for testing and evaluating survey questions", *Public Opinion Quarterly*, Vol. 68 No. 1, pp. 109–130.

Raineri, N. and Paillé, P. (2016), "Linking corporate policy and supervisory support with environmental citizenship behaviors: The role of employee environmental beliefs and commitment", *Journal of Business Ethics (in press)*.

Randall, M.L., Cropanzano, R., Bormann, C.A. and Birjulin, A. (1999), "Organizational politics and organizational support as predictors of work attitudes, job performance, and organizational citizenship behavior", *Journal of Organizational Behavior*, Vol. 20 No. 2, pp. 159–174.

Reid, S.E. and De Brentani, U. (2004), "The fuzzy front end of new product development for discontinuous innovations: A theoretical model", *Journal of Product Innovation Management*, Vol. 21 No. 3, pp. 170–184.

Reid, S.E. and De Brentani, U. (2012), "Organizational encouragement of divergent thinking, individual ideation behavior and divergent thinking attitudes: Impacts on market visioning competence", *Bishop's University Working Paper Series*, pp. 1–49.

Rhoades, L. and Eisenberger, R. (2002), "Perceived organizational support: A review of the literature", *Journal of Applied Psychology*, Vol. 87 No. 4, pp. 698–714.

Rhoades, L., Eisenberger, R. and Armeli, S. (2001), "Affective commitment to the organization: The contribution of perceived organizational support", *Journal of Applied Psychology*, Vol. 86 No. 5, pp. 825–836.

Riedl, C., Blohm, I., Leimeister, J.M. and Krcmar, H. (2010), "Rating scales for collective intelligence in innovation communities: Why quick and easy decision making does not get it right", *Proceedings of the International Conference on Information Systems*.

Roetzel, P.G. and Lohmann, C. (2014), "The influence of the perception of fairness on innovation idea value and knowledge sharing behavior in innovation idea networks", *Proceedings of the European Conference on Information Systems*.

Rohrbeck, R., Steinhoff, F. and Perder, F. (2008), "Virtual customer integration in the innovation process: Evaluation of the web platforms of multinational enterprises (MNE)", *PICMET 2008 Proceedings*.

Roldán, J.L. and Sánchez-Franco, M.J. (2012), "Variance-based structural equation modeling: Guidelines for using partial least squares", in Mora, M., Gelman, O., Steenkamp, A. and Raisinghani, M.S. (Eds.), *Research Methodologies, Innovations and Philosophies in Software Systems Engineering and Information Systems*, 1st ed., IGI Global, Hershey, pp. 193–221.

Roth, S., Schneckenberg, D. and Tsai, C.-W. (2015), "The ludic drive as innovation driver: Introduction to the gamification of innovation", *Creativity and Innovation Management*, Vol. 24 No. 2, pp. 300–306.

Rothwell, R. (1994), "Towards the fifth-generation innovation process", *International Marketing Review*, Vol. 11 No. 1, pp. 7–31.

Scheiner, C.W. (2015), "The Motivational Fabric of Gamified Idea Competitions: The Evaluation of Game Mechanics from a Longitudinal Perspective.", *Creativity and Innovation Management*, Vol. 24 No. 2, pp. 341–352.

Schlagwein, D., Schoder, D. and Fischbach, K. (2011), "Social information systems: Review, framework, and research agenda", *Proceedings of the International Conference on Information Systems*.

Schmelter, R., Mauer, R., Engelen, A. and Brettel, M. (2010), "Conjuring the entrepreneurial spirit in small and medium-sized enterprises: The influence of management on corporate entrepreneurship", *International Journal of Entrepreneurial Venturing*, Vol. 2 No. 2, pp. 159–184.

Schmidt, K., Heath, C. and Rodden, T. (2002), "Preface [Special issue on 'Awareness in CSCW']", *Computer Supported Cooperative Work*, Vol. 11 No. 3-4, pp. iii–iv.

Schulze, T., Indulska, M., Geiger, D. and Korthaus, A. (2012a), "Idea assessment in open innovation: A state of practice", *Proceedings of the European Conference on Information Systems*.

Schulze, T., Krug, S. and Schader, M. (2012b), "Workers' task choice in crowdsourcing and human computation markets", *Proceedings of the International Conference on Information Systems*.

Scott, S.G. and Bruce, R.A. (1994), "Determinants of innovative behavior: A path model of individual innovation in the workplace", *The Academy of Management Journal*, Vol. 37 No. 3, pp. 580–607.

Seawright, J. and Gerring, J. (2008), "Case selection techniques in case study research: A menu of qualitative and quantitative options", *Political Research Quarterly*, Vol. 61 No. 2, pp. 294–308.

Shanock, L.R. and Eisenberger, R. (2006), "When supervisors feel supported: Relationships with subordinates' perceived supervisor support, perceived organizational support, and performance", *Journal of Applied Psychology*, Vol. 91 No. 3, pp. 689–695.

Shenton, A.K. (2004), "Strategies for ensuring trustworthiness in qualitative research projects", *Education for Information*, Vol. 22 No. 2, pp. 63–75.

Shore, L.M. and Wayne, S.J. (1993), "Commitment and employee behavior: Comparison of affective commitment and continuance commitment with perceived organizational support", *Journal of Applied Psychology*, Vol. 78 No. 5, pp. 774–780.

Sloane, P. (2011), *A guide to open innovation and crowdsourcing: Advice from leading experts,* 1st ed., Kogan Page Publishers, London.

Smith, C.A., Organ, D.W. and Near, J.P. (1983), "Organizational citizenship behavior: Its nature and antecedents", *Journal of Applied Psychology*, Vol. 68 No. 4, pp. 653–663.

Somers, T.M. and Nelson, K. (2001), "The impact of critical success factors across the stages of enterprise resource planning implementations", *Proceedings of the Annual Hawaii International Conference on System Sciences.*

Spiller, S.A., Fitzsimons, G.J., Lynch, J. G. Jr. and McClelland, G.H. (2013), "Spotlights, floodlights, and the magic number zero: Simple effects tests in moderated regression", *Journal of Marketing Research*, Vol. 50 No. 2, pp. 277–288.

Spreitzer, G.M. (1995), "Psychological empowerment in the workplace: Dimensions, measurement, and validation", *Academy of Management Journal*, Vol. 38 No. 5, pp. 1442–1465.

Stol, K.-J. and Fitzgerald, B. (2015), "Inner source-adopting open source development practices in organizations: A tutorial", *IEEE Software*, Vol. 32 No. 4, pp. 60–67.

Strauss, A.L. (1987), *Qualitative analysis for social scientists,* 1st ed., Cambridge University Press, Cambridge.

Strauss, A.L. and Corbin, J. (1990), *Basics of qualitative research: Grounded theory procedures and techniques,* 15th ed., Sage Publications, Newbury Park.

Suh, T. and Shin, H. (2005), "Creativity, job performance and their correlates: A comparison between nonprofit and profit-driven organizations", *International Journal of Nonprofit and Voluntary Sector Marketing*, Vol. 10 No. 4, pp. 203–211.

Suh, T. and Shin, H. (2008), "When working hard pays off: Testing creativity hypotheses", *Corporate Communications: An International Journal*, Vol. 13 No. 4, pp. 407–417.

Tang, H.K. (1998), "An Integrative model of innovation in organizations", *Technovation*, Vol. 18 No. 5, pp. 297–309.

Terwiesch, C. and Xu, Y. (2008), "Innovation contests, open innovation, and multi-agent problem solving", *Management Science*, Vol. 54 No. 9, pp. 1529–1543.

Tesluk, P.E., Farr, J.L. and Klein, S.R. (1997), "Influences of organizational culture and climate on individual creativity", *The Journal of Creative Behavior*, Vol. 31 No. 1, pp. 27–41.

Thia, C.W., Chai, K.H., Bauly, J. and Xin, Y. (2005), "An exploratory study of the use of quality tools and techniques in product development", *The TQM Magazine*, Vol. 17 No. 5, pp. 406–424.

Thongpapanl, N. (2012), "The changing landscape of technology and innovation management: An updated ranking of journals in the field", *Technovation*, Vol. 32 No. 5, pp. 257–271.

Von Kortzfleisch, H., Hoeber, B. and Zerwas, D. (2015), "Innovationsmanagement", in Faltin, G. (Ed.), *Handbuch Entrepreneurship*, 1st ed., Springer, Wiesbaden, pp. 1–13.

Von Kortzfleisch, H., Mergel, I., Manoucheri, S. and Schaarschmidt, M. (2008), "Corporate web 2.0 applications: Motives, organisational embeddedness, and creativity", in Hass, B., Walsh, G. and Kilian, T. (Eds.), *Web 2.0: Neue Perspektiven für Marketing und Medien*, 1st ed., Springer, Berlin, Heidelberg, pp. 73–87.

Vujovic, S. and Ulhoi, J.P. (2008), "Opening up the innovation process", in Burton, R.M. (Ed.), *Designing organizations: 21st century approaches, Information and Organization Design Series*, Springer, New York.

Vukovic, M. and Bartolini, C. (2010), "Towards a research agenda for enterprise crowdsourcing", in Margaria, T. and Steffen, B. (Eds.), *Leveraging applications of formal methods, verification, and validation*, Springer, Berlin, Heidelberg, pp. 425–434.

Walsh, G., Yang, Z., Dose, D. and Hille, P. (2015), "The effect of job-related demands and resources on service employees' willingness to report complaints: Germany versus China", *Journal of Service Research*, Vol. 18 No. 2, pp. 193–209.

Wang, C.-J. and Tsai, C.-Y. (2014), "Managing innovation and creativity in organizations: An empirical study of service industries in Taiwan", *Service Business*, Vol. 8 No. 2, pp. 313–335.

Wasko, M.M. and Faraj, S. (2005), "Why should I share? Examining social capital and knowledge contribution in electronic networks of practice", *MIS Quarterly*, Vol. 29 No. 1, pp. 35–57.

Weiber, R. and Muehlhaus, D. (2014), *Strukturgleichungsmodellierung: Eine anwendungsorientierte Einführung in die Kausalanalyse mit Hilfe von AMOS, SmartPLS und SPSS*, 2nd ed., Springer, Wiesbaden.

Wernerfelt, B. (1984), "A resource-based view of the firm", *Strategic Management Journal*, Vol. 5 No. 2, pp. 171–180.

Wikhamn, B.R. and Knights, D. (2011), "Transaction cost economics and open innovation: Reinventing the wheel of boundary", *Proceedings of the DRUID Conference*.

Williams, S. and Schubert, P. (2015), *Social Business Readiness Studie 2014: Koblenz: CEIR Forschungsbericht*, Nr. 01/2015, Universität Koblenz-Landau.

Wooten, J.O. and Ulrich, K.T. (2015), "The impact of visibility in innovation tournaments: Evidence from field experiments", *SSRN Working paper*, No. 2214952.

Yang, Y., Chen, P.-Y. and Pavlou, P. (2009), "Open innovation: An empirical study of online contests", *Proceedings of the International Conference on Information Systems*.

Yeh-Yun Lin, C. and Liu, F. (2012), "A cross-level analysis of organizational creativity climate and perceived innovation", *European Journal of Innovation Management*, Vol. 15 No. 1, pp. 55–76.

Yousef, D.A. (2000), "Organizational commitment: A mediator of the relationships of leadership behavior with job satisfaction and performance in a non-western country", *Journal of Managerial Psychology*, Vol. 15 No. 1, pp. 6–24.

Yu, L. and Nickerson, J.A. (2011), "Generating creative ideas through crowds: An experimental study of combination", *Proceedings of the International Conference on Information Systems*.

Yuecesan, E. (2013), "An efficient ranking and selection approach to boost the effectiveness of innovation contests", *IEEE Transactions*, Vol. 45 No. 7, pp. 751–762.

Zait, P.A. and Bertea, P.E. (2011), "Methods for testing discriminant validity", *Management & Marketing Journal*, Vol. 9 No. 2, pp. 217–224.

Zhang, X. and Bartol, K.M. (2010), "Linking empowering leadership and employee creativity: The influence of psychological empowerment, intrinsic motivation, and creative process engagement", *Academy of Management Journal*, Vol. 53 No. 1, pp. 107–128.

Zhao, X., Lynch, J.G. and Chen, Q. (2010), "Reconsidering Baron and Kenny: Myths and truths about mediation analysis", *Journal of Consumer Research*, Vol. 37 No. 2, pp. 197–206.

Zheng, H., Li, D. and Hou, W. (2011), "Task design, motivation, and participation in crowdsourcing contests", *International Journal of Electronic Commerce*, Vol. 15 No. 4, pp. 57–88.

Zheng, H., Xie, Z., Hou, W. and Li, D. (2014), "Antecedents of solution quality in crowdsourcing: The sponsor's perspective", *Journal of Electronic Commerce Research*, Vol. 15 No. 3, pp. 212–224.

Zhou, J. and Shalley, C.E. (2003), "Research on employee creativity: A critical review and directions for future research", *Research in Personnel and Human Resources Management*, Vol. 22 No. 1, pp. 165–218.

Zhou, T. (2011), "Understanding online community user participation: A social influence perspective", *Internet Research*, Vol. 21 No. 1, pp. 67–81.

Zizlavsky, O. (2013), "Past, present and future of the innovation process", *International Journal of Engineering Business Management*, Vol. 5 No. 47, pp. 1–8.

Appendices

Appendix A: Results of the exploratory factor analysis (Study A)

In the following, the results of the exploratory factor analysis as conducted in Study A are detailed (see Table 62 for *organizational encouragement*, see Table 63 for *supervisory encouragement*, see Table 64 for *organizational impediments*, see Table 65 for *workload pressure*, see Table 66 for *affective organizational commitment*, see Table 67 for *motivation*, and see Table 68 for *participation intention*).

Organizational encouragement:

Indicator	MSA	Commonality	Factor 1	Factor 2
Enthusiasm	0.954	0.634	**0.643[1]**	0.208[1]
Idea evaluation	0.937	0.590	**0.688**	0.114
Failure acceptance	0.926	0.533	**0.887**	-0.285
Performance evaluation	0.909	0.574	**0.782**	-0.038
Atmosphere	0.897	0.555	**0.794**	-0.076
Active idea flow	0.926	0.603	**0.637**	0.190
Shared vision	0.938	0.527	**0.433**	0.362
Problem solving	0.913	0.648	-0.049	**0.837**
Creative work	0.872	0.647	-0.352	**0.994**
Support of new ideas	0.949	0.662	0.362	**0.527**
Recognition	0.931	0.594	0.333	**0.509**
Rewarding creativity	0.897	0.439*	0.030	0.642
Mechanisms	0.946	0.493*	0.202	0.551
Risk taking	0.903	0.197*	0.440	0.005
Handling unusual ideas	0.923	0.308*	0.280	0.328

Note: MSA= Measure of Sampling Adequacy, Rotation: Promax, [1]Interpretation: Pattern matrix
*Indicators excluded because of low commonality values (< 0.5)

Table 62: Exploratory factor analysis for *organizational encouragement*[102]

---

[102] Author's own table.

Supervisory encouragement:

Indicator	MSA	Commonality	Factor 1
Expectations	0.895	0.689	**0.830[1]**
Planning	0.951	0.635	**0.797**
Goal setting	0.897	0.602	**0.776**
Communication	0.932	0.728	**0.853**
Interpersonal skills	0.928	0.681	**0.825**
Confidence	0.901	0.683	**0.826**
Valuing contributions	0.935	0.695	**0.834**
Serves as good example	0.933	0.789	**0.888**

Note: MSA= Measure of Sampling Adequacy, Rotation: Promax, [1]Interpretation: Pattern matrix

Table 63: Exploratory factor analysis for *supervisory encouragement*[103]

Organizational impediments:

Indicator	MSA	Commonality	Factor 1	Factor 2
Idea criticism	0.917	0.506	**0.557[1]**	0.221[1]
Work criticism	0.917	0.549	**0.669**	0.112
Work pressure	0.901	0.516	**0.781**	-0.116
Formal procedures	0.883	0.582	**0.768**	-0.009
Strict control	0.831	0.504	**0.827**	-0.246
Political problems	0.896	0.666	-0.016	**0.826**
Destructive competition	0.867	0.645	0.047	**0.774**
Project hindering	0.895	0.646	0.016	**0.794**
Destructive criticism	0.926	0.597	0.270	**0.581**
Protecting territories	0.906	0.635	0.398	**0.494**
TMT risk taking	0.582	0.250*	-0.410	0.617
Change emphasis	0.868	0.351*	0.607	-0.026

Note: MSA= Measure of Sampling Adequacy, Rotation: Promax, [1]Interpretation: Pattern matrix
*Indicators excluded because of low commonality values (< 0.5)

Table 64: Exploratory factor analysis for *organizational impediments*[104]

---

[103] Author's own table.
[104] Author's own table.

Workload pressure:

Indicator	MSA	Commonality	Factor 1	Factor 2
Too much work	0.730	0.701	**0.811[1]**	0.075[1]
Distractions	0.775	0.656	**0.845**	-0.165
Time pressure	0.719	0.731	**0.830**	0.071
Sufficient time (r)	0.704	0.689	0.253	**0.717**
Realistic expectations (r)	0.577	0.801	-0.183	**0.934**

Note: MSA= Measure of Sampling Adequacy, Rotation: Promax, [1]Interpretation: Pattern matrix

Table 65: Exploratory factor analysis for *workload pressure*[105]

Affective organizational commitment:

Indicator	MSA	Commonality	Factor 1	Factor 2
Rest of career	0.931	0.617	**0.786[1]**	-0.018[1]
Organization's problems	0.916	0.586	**0.675**	-0.368
Part of the family	0.861	0.654	**0.801**	0.103
Emotionally attached	0.863	0.814	**0.902**	0.026
Personal meaning	0.866	0.736	**0.830**	0.208
Sense of belonging	0.859	0.802	**0.884**	0.134
Discussing with outside people	0.885	0.470*	0.594	-0.348
Attachment to other firms	0.450**	0.819	0.120	0.896

Note: MSA= Measure of Sampling Adequacy, Rotation: Promax, [1]Interpretation: Pattern matrix
*Indicators excluded because of low commonality values (< 0.5)
**Indicators excluded because of low MSA values (< 0.5)

Table 66: Exploratory factor analysis for *affective organizational commitment*[106]

---

[105] Author's own table.
[106] Author's own table.

Motivation:

Indicator	MSA	Commonality	Factor 1
Curiosity	*0.853*	*0.660*	***0.812[1]***
Challenging oneself	*0.729*	*0.687*	***0.829***
Figuring out own Proficiency	*0.742*	*0.754*	***0.868***
Making new experiences	*0.816*	*0.488**	*0.698*
Like the things I do	*0.910*	*0.490**	*0.700*

Note: MSA= Measure of Sampling Adequacy, Rotation: Promax, [1]Interpretation: Pattern matrix

Table 67: Exploratory factor analysis for *motivation*[107]

Participation intention:

Indicator	MSA	Commonality	Factor 1
Intention to participate	*0.733*	*0.918*	***0.958[1]***
Attempt to participate	*0.742*	*0.914*	***0.956***
Decision to participate	*0.835*	*0.880*	***0.938***

Note: MSA= Measure of Sampling Adequacy, Rotation: Promax,[1]Interpretation: Pattern matrix

Table 68: Exploratory factor analysis for *participation intention*[108]

---

[107] Author's own table.
[108] Author's own table.

Appendix B: Results of the confirmatory factor analysis (Study A)

In the following, the results of the confirmatory factor analysis as conducted in Study A are detailed (see Table 69 and Table 70 for *organizational encouragement*, see Table 71 for *supervisory encouragement*, see Table 72 and Table 73 for *organizational impediments*, see Table 74 for *workload pressure*, see Table 75 for *affective organizational commitment*, see Table 76 for *motivation*, and see Table 77 for *participation intention*).

Organizational encouragement:

Org. encourage-ment Indicators	Confirmatory factor analysis			Reliabilities	
	Factor loadings	Loadings square	Error variance	Indicator reliability	Factor reliability
Enthusiasm	*0.757*	*0.573*	*0.427*	*0.573*	
Performance evaluation	*0.712*	*0.507*	*0.493*	*0.507*	
Atmosphere	*0.713*	*0.508*	*0.492*	*0.508*	
Idea flow	*0.765*	*0.585*	*0.415*	*0.585*	*0.872*
Shared vision	*0.707*	*0.500*	*0.500*	*0.500*	
Idea evaluation	*0.706*	*0.498*	*0.502*	*0.498**	
Failure acceptance	*0.540*	*0.292*	*0.708*	*0.292**	

Note: *Indicator(s) excluded from further analysis because of low indicator reliability (< 0.5)

Table 69: Confirmatory factor analysis for *organizational encouragement (factor 1)*[109]

---

[109] Author's own table.

Org. encourage-ment Indicators	Confirmatory factor analysis			Reliabilities	
	Factor loadings	Loadings square	Error variance	Indicator reliability	Factor reliability
Support of new ideas	0.817	0.667	0.333	0.667	
Recognition	0.748	0.559	0.441	0.559	0.797
Expectations	0.539	0.291	0.709	0.291*	
Problem solving	0.698	0.488	0.512	0.488*	

Note: *Indicator(s) excluded from further analysis because of low indicator reliability (< 0.5)

Table 70: Confirmatory factor analysis for *organizational encouragement (factor 2)*[110]

Supervisory encouragement:

Sup. encourage-Indicators	Confirmatory factor analysis			Reliabilities	
	Factor loadings	Loadings square	Error variance	Indicator reliability	Factor reliability
Expectations	0.785	0.616	0.384	0.616	
Planning	0.754	0.569	0.431	0.569	
Goal setting	0.729	0.531	0.469	0.531	
Communication	0.840	0.705	0.295	0.705	0.935
Interpersonal skills	0.804	0.646	0.354	0.646	
Confidence	0.805	0.647	0.353	0.647	
Valuing contributions	0.811	0.658	0.342	0.658	
Serves as good example	0.882	0.778	0.222	0.778	

Table 71: Confirmatory factor analysis for *supervisory encouragement*[111]

---

[110] Author's own table.
[111] Author's own table.

Organizational impediments:

Org. impediments (FACTOR 1) Indicators	Confirmatory factor analysis			Reliabilities	
	Factor loadings	Loadings square	Error variance	Indicator reliability	Factor reliability
Work criticism	*0.742*	*0.551*	*0.449*	*0.551*	
Idea criticism	*0.691*	*0.477*	*0.523*	*0.477*	*0.766*
Work pressure	*0.548*	*0.433*	*0.567*	*0.433**	
Formal procedures	*0.650*	*0.422*	*0.578*	*0.422**	

Note: *Indicator(s) excluded from further analysis because of low indicator reliability (< 0.5)

Table 72: Confirmatory factor analysis for *organizational impediments (factor 1)*[112]

Org. impediments (FACTOR 2) Indicators	Confirmatory factor analysis			Reliabilities	
	Factor loadings	Loadings square	Error variance	Indicator reliability	Factor reliability
Political problems	*0.736*	*0.542*	*0.458*	*0.542*	
Destructive competition	*0.768*	*0.589*	*0.411*	*0.589*	
Protecting territories	*0.751*	*0.563*	*0.437*	*0.563*	*0.861*
Project hindering	*0.728*	*0.530*	*0.470*	*0.530*	
Destructive criticism	*0.737*	*0.544*	*0.456*	*0.544*	

Table 73: Confirmatory factor analysis for *organizational impediments (factor 1)*[113]

---

[112] Author's own table.
[113] Author's own table.

Workload pressure:

Workload pressure (FACTOR 1)	Confirmatory factor analysis			Reliabilities	
Indicators	Factor loadings	Loadings square	Error variance	Indicator reliability	Factor reliability
Workload	0.728	0.530	0.470	0.530	
Time pressure	0.803	0.644	0.356	0.644	*0.782*
Distractions	0.681	0.464	0.536	0.464*	

Note: *Indicator(s) excluded from further analysis because of low indicator reliability (< 0.5)

Table 74: Confirmatory factor analysis for *workload pressure (factor 1)*[114]

Affective organizational commitment:

Affect. org. com-	Confirmatory factor analysis			Reliabilities	
Indicators	Factor loadings	Loadings square	Error variance	Indicator reliability	Factor reliability
Spend rest of career	0.708	0.501	0.499	0.501	
Part of the family	0.793	0.629	0.371	0.629	
Emotionally attached	0.901	0.812	0.188	0.812	*0.913*
Personal meaning	0.806	0.650	0.350	0.650	
Sense of belonging	0.896	0.803	0.197	0.803	

Table 75: Confirmatory factor analysis for *affective organizational commitment*[115]

Motivation:

Motivation	Confirmatory factor analysis			Reliabilities	
Indicators	Factor loadings	Loadings square	Error variance	Indicator reliability	Factor reliability
Challenging oneself	0.890	0.792	0.208	0.792	*0.879*
Figuring out own profi-	0.881	0.777	0.223	0.777	

Table 76: Confirmatory factor analysis for *motivation*[116]

---

[114] Author's own table.
[115] Author's own table.
[116] Author's own table.

Participation intention:

Participation inten- Indicators	Confirmatory factor analysis			Reliabilities	
	Factor loadings	Loadings square	Error variance	Indicator reliability	Factor reliability
Intention to participate	*0.938*	*0.881*	*0.119*	*0.881*	
Attempt to participate	*0.941*	*0.885*	*0.115*	*0.885*	***0.947***
Decision to participate	*0.898*	*0.806*	*0.194*	*0.806*	

Table 77: Confirmatory factor analysis for *participation intention*[117]

Results of renewed confirmatory factor analysis:

Scales	Factor reliability	# of in-dicators	Indicators
Organizational encouragement	*0.890*	*5*	*Enthusiasm, Idea flow, Shared vision, Support of new Ideas, Recogni-tion*
Supervisory encouragement	*0.935*	*8*	*Expectations, Planning, Goal set-ting, Communication, Interpersonal skills, Confidence, Valuing contri-butions, Serves as good example*
Organizational impediments	*0.874*	*4*	*Destructive competition, Protecting territories, Project hindering, De-structive criticism*
Workload pressure	*0.768*	*2*	*Workload, Time pressure*
Affective or-ganizational commitment	*0.913*	*5*	*Spend rest of career, Part of the fam-ily, Emotionally attached, Personal meaning, Sense of belonging*
Motivation	*0.879*	*2*	*Challenging oneself, Figuring out own proficiency*
Participation intention	*0.947*	*3*	*Intention to participate, Attempt to participate, Decision to participate*

Table 78: Results of renewed confirmatory factor analysis[118]

---

[117] Author's own table.
[118] Author's own table.

Appendix C: Indicators to measure latent variables (Study A)

In the following, the items used for measuring the latent constructs are detailed (see Table 79 for *organizational encouragement*, see Table 80 for *supervisory encouragement*, see Table 81 for *organizational impediments*, see Table 82 for *workload pressure*, see Table 83 for *affective organizational commitment*, see Table 84 for *motivation*, and see Table 85 for *participation intention*).

Organizational encouragement:

ID	Indicator	Statements
OE_1	Problem solving	People are encouraged to solve problems creatively in this organization
OE_2	Support of new ideas	New ideas are encouraged in this organization
OE_3	Mechanisms	This organization has a good mechanism for encouraging and developing creative ideas
OE_4	Risk taking	People are encouraged to take risks in this organization
OE_5	Expectations	In this organization, top management expects that people will do creative work
OE_6	Enthusiasm	I feel that top management is enthusiastic about my project(s)
OE_7	Idea evaluation	Ideas are judged fairly in this organization
OE_8	Handling unusual ideas	People in this organization can express unusual ideas without the fear of being called stupid
OE_9	Failure acceptance	Failure is acceptable in this organization, if the effort on the project was good
OE_10	Performance evaluation	Performance evaluation in this organization is fair
OE_11	Recognition	People are recognized for creative work in this organization
OE_12	Rewarding creativity	People are rewarded for creative work in this organization
OE_13	Atmosphere	There is an open atmosphere in this organization
OE_14	Idea flow	In this organization, there is a lively and active flow of ideas
OE_15	Shared vision	Overall, the people in this organization have a shared vision of where we are going and what we are trying to do

Table 79: Indicators for *organizational encouragement*[119]

---

[119] Author's own table, adapted from Amabile (2010).

Supervisory encouragement:

ID	Indicator	Statements
SE_1	Expectations	My supervisor's expectations for my project(s) are clear
SE_2	Planning	My supervisor plans well
SE_3	Goal setting	My supervisor clearly sets overall goals for me
SE_4	Communication	My supervisor communicates well with our work group
SE_5	Interpersonal skills	My supervisor has good interpersonal skills
SE_6	Confidence	My supervisor shows confidence in our work group
SE_7	Valuing contributions	My supervisor values individual contributions to project(s)
SE_8	Serves as good example	My supervisor serves as a good work model

Table 80: Indicators for *supervisory encouragement*[120]

Organizational impediments:

ID	Indicator	Statements
OI_1	Political problems	There are political problems in this organization
OI_2	Destructive competition	There is destructive competition within this organization
OI_3	Protecting territories	People in this organization are very concerned about protecting their territory
OI_4	Project hindering	Other areas of the organization do hinder my project(s)
OI_5	Idea criticism	People are critical of new ideas in this organization
OI_6	Destructive criticism	Destructive criticism is a problem in this organization
OI_7	Work criticism	People are concerned about negative criticism of their work in this organization
OI_8	Work pressure	People in this organization do feel pressure to produce anything acceptable, even if quality is lacking
OI_9	TMT risk taking	Top management is willing to take risks in this organization
OI_10	Change emphasis	There is emphasis in this organization on doing things the way we have always done them
OI_11	Formal procedures	Procedures and structures are too formal in this organization
OI_12	Strict control	This organization is strictly controlled by the upper management

Table 81: Indicators for *organizational impediments*[121]

---

[120] Author's own table, adapted from Amabile (2010).
[121] Author's own table, adapted from Amabile (2010).

Workload pressure:

ID	Indicator	Statements
WP_1	Workload	I do have too much work to do in too little time
WP_2	Sufficient time$^{(r)}$	I have sufficient time to do my project(s)
WP_3	Distractions	There are too many distractions from project work in this organization
WP_4	Realistic expectations$^{(r)}$	There are realistic expectations for what people can achieve in this organization
WP_5	Time pressure	I do feel a sense of time pressure in my work
(r)= Reverse-coded		

Table 82: Indicators for *workload pressure*[122]

Affective organizational commitment:

	Indicator	Statements
AC_1	Spend rest of career	I would be very happy to spend the rest of my career with this organization
AC_2	Discussing with outside people	I enjoy discussing my organization with people outside it
AC_3	Organization's problems	I really feel as if this organization's problems are my own
AC_4	Attachment to other firms$^{(r)}$	I think that I could easily become as attached to another organization as I am to this one
AC_5	Part of the family	I do feel like 'part of the family' at my organization
AC_6	Emotionally attached	I do feel 'emotionally attached' to this organization
AC_7	Personal meaning	This organization has a great deal of personal meaning for me
AC_8	Sense of belonging	I do feel a strong sense of belonging to my organization
(r)= Reverse-coded		

Table 83: Indicators for *affective organizational commitment*[123]

---

[122] Author's own table, adapted from Amabile (2010).
[123] Author's own table, adapted from Allen and Meyer (1990).

Motivation:

ID	Indicator	Statements
M_1	Making new experiences	No matter what the outcome of the contest, I am satisfied if I gained a new experience
M_2	Like the things I do	What matters most to me is enjoying what I do in this Innovation Contest
M_3	Curiosity	Curiosity is the driving force behind much of what I do in this Innovation Contests
M_4	Challenging oneself	I want to challenge myself to solve the problem in this Innovation Contest
M_5	Figuring out own proficiency	I want to find out how good I really can be at this Innovation Contest

Table 84: Indicators for *motivation*[124]

Participation intention:

ID	Indicator	Statements
PI_1	Intention to participate	I intend to participate in this Innovation Contest
PI_2	Attempt to participate	I will try to participate in this Innovation Contest
PI_3	Decision to participate	I am determined to participate in this Innovation Contest

Table 85: Indicators for *participation intention*[125]

---

[124] Author's own table, adapted from Zheng *et al.* (2011).
[125] Author's own table, adapted from Zheng *et al.* (2011).

Appendix D: Final coding scheme for the content analysis (Study B)

The following figures depict the coding schemes for the strategic level (see Figure 24), the initiative level (see Figure 25) and the contest level (see Figure 26).

Strategic Level:

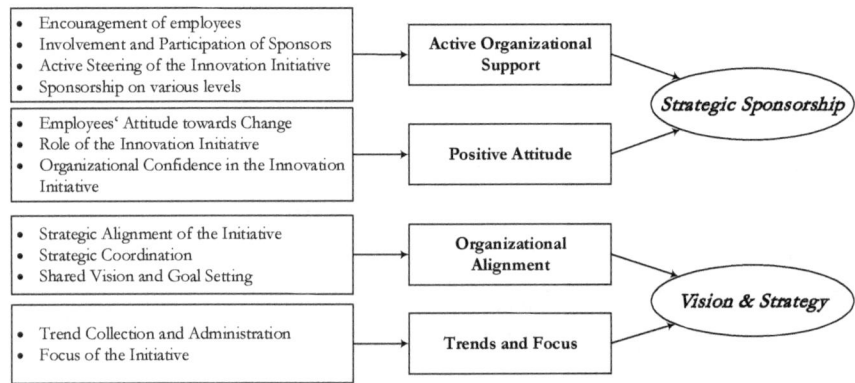

Figure 24: Coding scheme (strategic level)[126]

Initiative Level:

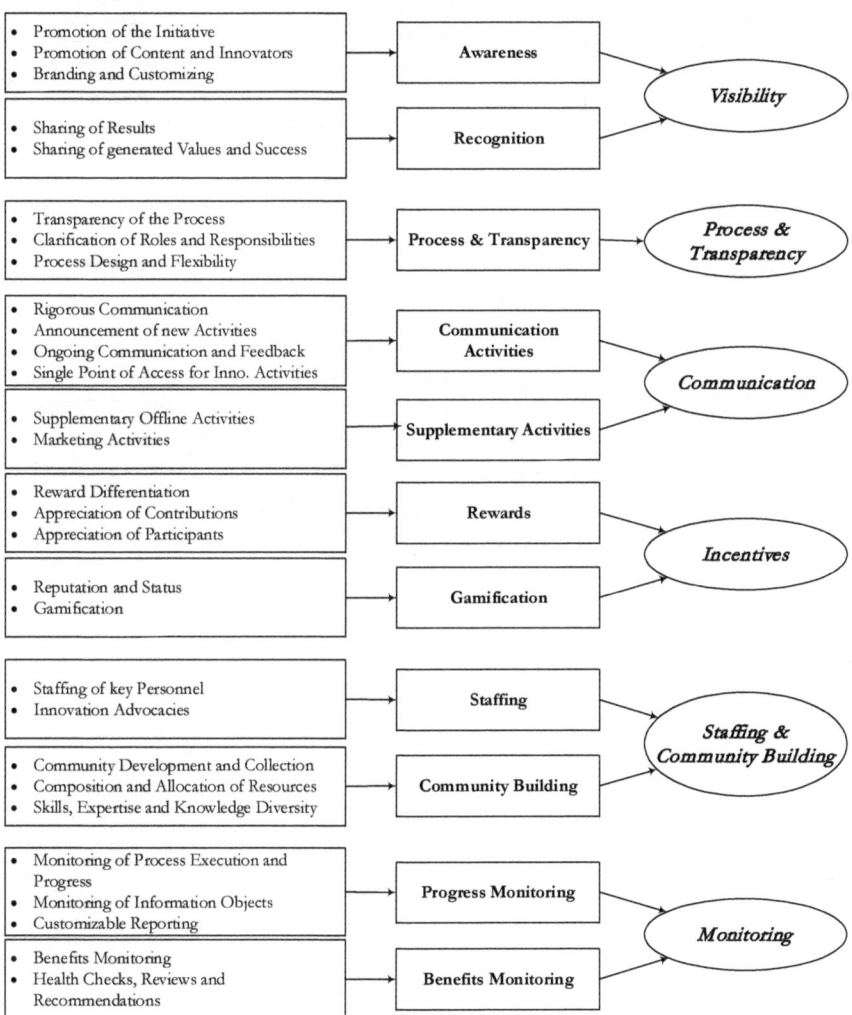

Figure 25: Coding scheme (initiative level)[127]

Contest Level:

[127] Author's own figure.

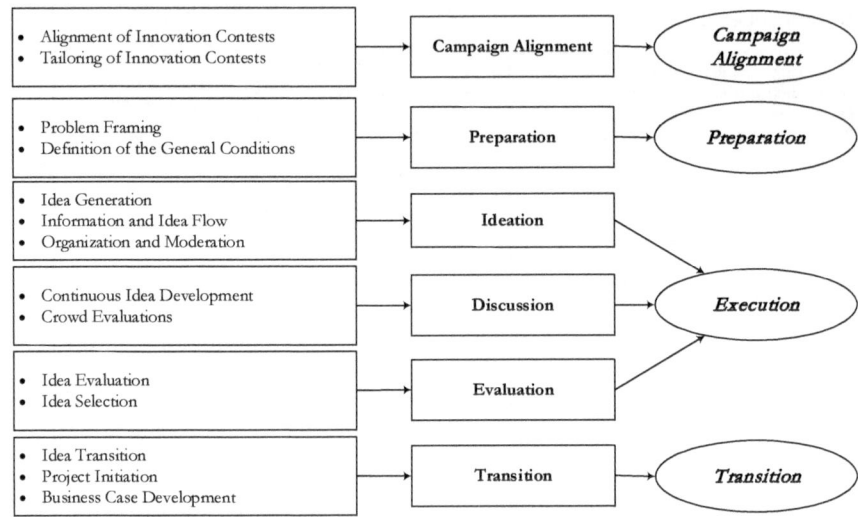

Figure 26: Coding scheme (contest level)[128]

---

[128] Author's own figure.